Storey's Guide to

RAISING DAIRY GOATS

Jerry Belanger

Storey Publishing

*The mission of Storey Publishing is to serve our customers
by publishing practical information that encourages personal independence
in harmony with the environment.*

Edited by Marie Salter
Cover design by Renelle Moser
Front cover photograph © John Colwell from Grant Heilman
Back cover photograph from PhotoDisc
Illustrations by Elayne Sears, except those on pages 18, 21, and 85 by Chuck Galey;
67 and 69 (top) by Carl Kirkpatrick; 87 (chicory, daisy, nettle, yarrow) and 88
(mountain laurel, oak leaf, wild cherry) by Beverly Duncan; 87 (dandelion, this-
tle) by Louise Riotte; 87 (multiflora rose) by Sarah Brill; 87 (plantain) and 106 by
Alison Kolesar; 88 (bracken fern, locoweed, milkweed, sorrel) and 89 (amaranth)
by Regina Hughes; 88 (dock) by Judy Eliason; 88 (rhubarb) by Mallory Lake; 89
(Johnson grass) by Bobbi Angell; 111 and 176 by Brigita Fuhrmann; 186–189
based on plans provided by *Countryside & Small Stock Journal.*
Series design by Mark Tomasi
Text production by Susan Bernier and Deborah Daly
Indexed by Nan Badgett/Word•a•bil•i•ty

The information in this book is true and complete to the best of our knowledge. All recommendations are made without guarantee on the part of the author or Storey Publishing. The
author and publisher disclaim any liability in connection with the use of this information. For
additional information please contact Storey Publishing, 210 MASS MoCA Way, North
Adams, MA 01247.
Storey books are available for special premium and promotional uses and for customized edi-
tions. For further information, please call 1-800-793-9396.

Storey's Guide to Raising Dairy Goats was previously published under the title *Raising Milk Goats
the Modern Way.* This new edition has been expanded by 80 pages. All of the information in
the previous edition was reviewed and revised for this new text, which offers the most compre-
hensive and up-to-date information available on the topic of raising dairy goats.

Printed in the United States by Versa Press
20 19 18 17 16 15 14 13 12

Library of Congress Cataloging-in-Publication Data

Belanger, Jerome D.
    Storey's guide to raising dairy goats / Jerry Belanger.
      p. cm.
    Includes index.
    ISBN-13: 978-1-58017-259-2
    ISBN-10: 1-58017-259-8
    1. Goats. I. Title: Guide to raising dairy goats. II. Title.
SF383 .B45 2000
636.3'9142—dc21
                                                           00-057389

CONTENTS

BASIC INFORMATION ABOUT GOATS

This book assumes that you are interested in goats, and that you like these interesting and valuable animals, but it doesn't assume that you know anything about them. So let's start at the very beginning by looking at some basic terms and facts. (If you already know the basics, or if you're more interested in practical matters than in terminology and history, feel free to skip ahead to the next chapter.)

Basic Terms

Female goats are called *does*, or if they're less than a year old, sometimes *doelings*. Males are *bucks*, or *bucklings*. Young goats are *kids*. In polite dairy goat company they are never *nannies* or *billies*, although you might hear these terms applied to meat goats. Correct terminology is important to those who are working to improve the image of the dairy goat. People who think of a *nanny goat* as a stupid and smelly beast that produces small amounts of vile milk will at least have to stop to consider the truth if she's called a doe.

Gender Differences
and the Truth about Goat Aroma

Does are *not* smelly. They are not mean, and of course they don't eat tin cans. They are dainty, fastidious about what they eat, intelligent (smarter than dogs, some scientists tell us), friendly, and a great deal of fun to have around.

Bucks have two major scent glands located between and just to the rear of the horns or horn knobs, and minor ones in the neck region. Bucks do smell, but the does think it's great and some goat raisers don't mind it either. The odor is strongest during the breeding season, which usually runs from September to about January. The scent glands can be removed, although some authorities frown on the practice, because a descented buck can be less efficient at detecting and stimulating estrus and will still have an odor.

Still, even if they don't stink, bucks have habits that make them less than ideal family pets. For instance, they urinate all over their front legs and beards or faces. This is natural, but it tends to turn some people off.

In most cases the home dairy won't even have a buck (see chapter 9), so we return to the fact that you can keep goats even if you have neighbors or if your barn is fairly close to the house, and no one will be overpowered by goat aroma.

Livestock or Pets?

One of the challenges of goat public relations is that everyone seems to have had one in the past or knows someone who did. Most of them were pets, and that's where the trouble lies.

A goat is not much bigger than a large dog (average weight for a doe is less than 150 pounds), it's no harder to handle, and it does make a good pet. But a goat is not a dog. People who treat it like one are asking for trouble, and when they get rid of the poor beast in disgust, they bring trouble down on all goats and all goat lovers. If the goat "eats" the clothes off the line or nips off the rose bushes or the pine trees, strips the bark off young fruit trees, or butts people, it's not the goat's fault but the owner's.

Goats are livestock. Would you let a cow or a pig roam free and then damn the whole species when one got into trouble? Would you condemn all dogs because one is vicious, because it was chained, beaten, and teased? Children can have fun playing with goats, but when they "teach" a young kid to butt people and that kid grows up to be a 200-pound male who still wants to play, there's bound to be trouble. Likewise, a mistreated animal of any species isn't likely to have a docile disposition.

Because goats are livestock, and more specifically dairy animals, they must be treated as such. That means not only proper housing and feed, but strict attention to and regularity of care. If you can't or won't want to milk at 12-hour intervals, even when you're tired or under the weather; if the thought of staying home weekends and vacations depresses you and you can't count on the help of a friend or neighbor; then don't even consider raising goats. The rewards of goat raising are great and varied, but you don't get rewards without working for them.

Goats Eat Everything, Don't They?

The goat (*Capra hircus*) is related to the deer, not to dogs, cats, or even cows. It is a browser rather than a grazer, which means it would rather reach up than down for food. The goat also craves variety. Couple all that with its natural curiosity and nothing is safe from at least a trial taste. Anything hanging, like clothes on a wash line, is just too much for a goat's natural instincts to resist.

Rose bushes and pine trees are high in vitamin C and goats love them. Leaves, branches, and the bark of young trees are a natural part of the goat's diet in the wild. If you treat a browsing goat like a carnivorous dog, of course you'll have problems! But don't blame the goat.

Goat Housing

Like a cow or a pig, a goat needs a sturdily-fenced pen. Each doe requires a minimum of 10 square feet of inside space, plus as much outdoor space as you can manage. Goats do not require pasture, and unless it contains browse they probably won't utilize much of it anyway. They'll trample more than they eat. It's better to bring their food to them and feed them in a properly constructed manger, especially in a land- and labor-intensive small-farm situation. (See chapter 4 for more on housing.)

Goats are *not* lawn mowers. Most of them won't eat lawn grass, unless starved to it, and they won't produce milk on it.

Goats eat tin cans? Of course not. But they'll eat (or at least taste) the paper and glue on tin cans, which probably started the myth.

Important Note

Never stake out a goat. There is too much danger of strangulation, and many goats have been injured or killed by dogs. Even the family pet you thought was a friend of the goat could turn on it.

Goats can be raised in a relatively small area. If there are no zoning regulations restricting livestock, dairy goats can be (and are) raised even on average-sized lots in town.

A Little History

Goats have been humanity's companions and benefactors throughout recorded history, and even before. There is evidence that goats were among the first, some say *the* first, animals to be domesticated by humans, perhaps as long as 10,000 years ago. They provided meat, milk, skins, and undoubtedly entertainment and companionship.

Wild goats originated in Persia and Asia Minor (*Capra aegagrus)*, the Mediterranean basin (*Capra prisca),* and the Himalayas (*Capra falconeri).* There were domesticated goats *(Capra hircus)* in Switzerland by the middle period of the Stone Age. The first livestock registry in the world was organized in Switzerland in the 1600s — for goats.

Goats were distributed around the world by early explorers and voyagers. They were commonly carried on board ships as a source of milk and meat. There were goats, for example, aboard the *Mayflower* on its famous voyage to America in 1620. Thus, goats ended up on shores far from home and spread to most parts of the world. Many returned to their *feral,* or wild, state in their new homes. Today, there are an estimated 460 million goats in the world.

In Europe, goats provided more milk than cows did until well after the Middle Ages. Even today, in the world as a whole, more people use

The Biology of Goats

Goats are mammals, of the phylum Mammalia: their young are born alive, and suckle on a secretion from the mammary glands, which of course is milk.

They are of the order Artiodactyla, which means they are even-toed, hoofed mammals, and of the suborder Ruminantia, from the Latin meaning "to chew cud," and have four "stomachs" like cows.

They belong to the family Bovidae, which among other things means that they have hollow horns that they don't shed. (Some goats are naturally hornless, or *polled.* Many more are *disbudded:* the horn buds are burned out with a hot iron or with caustic before they start to grow. Some goats are *dehorned:* the horns grow but are then cut off. See chapter 7.)

Goats belong to the genus *Capra,* which includes only goats. We will discuss the species *Capra hircus,* the domestic goat. Within the species, subdivisions are known as *breeds.*

goat milk than cow milk. Goats are certainly more common in less fertile, or more arid, or developing countries than they are in the United States and Canada. They're more efficient animals than cattle are in their ability to convert plants into more valuable animal protein. Although goats are more labor intensive than cattle, this is of small concern in backyard dairies and nonindustrialized countries, and of no concern at all where there isn't enough feed for cattle to do well, or where a cow would produce more milk than a family could use.

Breeds of Goats

While all domestic goats have descended from a common parentage, there are many breeds, or subdivisions of the species, throughout the world — more than eighty. Only a few of these are found in the United States.

Goat breeds are classified according to their main purpose: that is, meat, mohair, or milk. In this book we'll be concentrating on the goats that have been bred for milk production, although in most respects care is the same for all.

Bear in mind that many, perhaps most, American goats are not pure-breds: they are mixed, and can't be identified as belonging to any particular breed. If these are fairly decent animals they're usually referred to as *grades*; if not, most people call them *scrubs*.

Nubian

The most popular pure breed in America is the Nubian. Nubians can be any color or color pattern, but they're easily recognized by their long drooping ears and Roman noses.

It's commonly said that the Nubian originated in Africa, but technically, the genealogy is a bit more complicated. From Africa, the Nubian made a stop along the way to the United States. Our Nubians are descendants of the Anglo-Nubian, which resulted from crossing native English goats with lop-eared breeds from Africa and India. The first three Nubians arrived in this country in 1909, imported by Dr. R. J. Gregg of Lakeside, California.

The Nubian is often compared with the Jersey of the cow world. The average Nubian produces less milk than the average goat of any other breed, but the average butterfat content is higher.

Averages can be misleading though. While the average production for a Nubian is about 1,600 pounds in 305 days, with 69 pounds of butterfat, U.S. Department of Agriculture (USDA) figures reveal one record of 5,940 pounds of milk and 303 pounds of fat.

Nubians are readily identified by their pendulous ears and Roman noses.

Saanen

Next in popularity behind the Nubian is the Saanen (pronounced SAH-nen). This is a pure white goat with erect ears and a "dished" face that is just the opposite of the Nubian's. Saanens originated in the Saane Valley of Switzerland, and have enjoyed a wider distribution throughout the world than any other breed. The first Saanens arrived in the United States in 1904.

They are large goats, with high average milk production: almost 1,900 pounds in 305 days. Butterfat averages 3.5 percent on a yearly basis. The all-time record is 6,571 pounds.

Until recently, Saanens that were not pure white or light cream were discriminated against in purebred circles. Any that were colored or spotted could not be registered, and they were frequently disposed of.

That changed in the 1980s when some Saanen breeders kept the colored or patterned animals, found that they were fine dairy animals, and started promoting them as a separate breed. They're not crossbreds: they're actually purebred Saanens, but with a "color defect" that results when both the sire and the dam carry a recessive color gene. Today, these goats are called *Sables*. Many promoters of Sables like to refer to their goats as "the Saanens in party clothes."

Saanens are always white and have "dished" or concave faces.

French Alpine

The French Alpine originated in the Alps and arrived in the United States in 1920, being imported by Dr. C. P. DeLangle. The color of Alpines varies greatly, ranging from white to black, and often one animal displays several colors and shades.

There are recognized color patterns, such as the *cou blanc* (French for "white neck"), a white neck and shoulders shade through silver gray to a glossy black on the hindquarters; there are gray or black markings on the head. Another color pattern, the *chamoisée*, can be tan, red, bay, or brown, with black markings on the head, a black stripe down the back, and black stripes on the hind legs. The *sundgau* has black and white markings on the face and underside. The *pied* is spotted or mottled; the *cou clair* has tan to white front quarters shading to gray, with black hindquarters; and the *cou noir* has black front quarters and white hindquarters.

According to 1998 USDA figures, Alpines have the highest annual average milk production of any breed, with 2,031 pounds and 3.5 percent butterfat. The record, set in 1997, is 6,414 pounds.

You might also hear of British Alpines, Rock Alpines (named not because they like to climb on rocks any more than other goats do, but because they were developed in America by Mary Edna Rock), and Swiss Alpines.

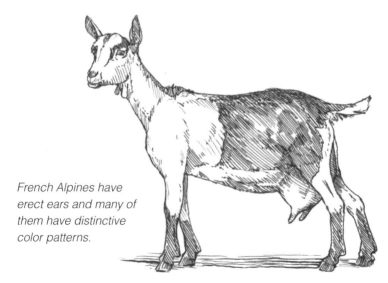

French Alpines have erect ears and many of them have distinctive color patterns.

Oberhasli

There are no more Swiss Alpines. No, they're not extinct. In 1978, their name was changed to Oberhasli *(Oh-ber-HAAS-lee)*. This goat was developed near Bern, Switzerland, where it is known as the Oberhasli-Brienzer, among other names.

The outstanding feature in the appearance of the Oberhasli is its rich red bay coat with black "trim." The black includes stripes down the face, ears, back, belly, and udder. The legs are also black below the knees and hocks. Oberhasli milk production averages 1,605 pounds of milk with 3.6 percent butterfat. The record is 4,655 pounds.

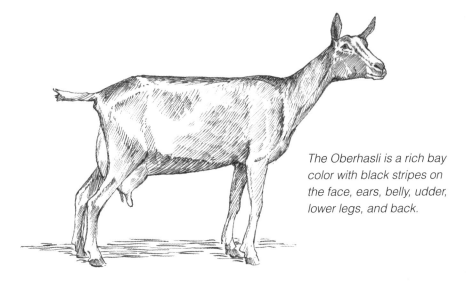

The Oberhasli is a rich bay color with black stripes on the face, ears, belly, udder, lower legs, and back.

How Many Goats?

According to the 1998 agricultural census, there are about 125,000 dairy goats in the United States. The leading states are California, Texas, Wisconsin, New York, and Ohio. However, the census only covers farmers (that is, those meeting minimum income requirements) and most homestead and backyard animals aren't included in these totals.

In last place? Hawaii, Delaware, Rhode Island, and Alaska.

Toggenburg

Toggenburgs are the oldest registered breed of any animal in the world, with a herd book having been established in Switzerland in the 1600s. They were the first imported purebreds to arrive in the United States, in 1893, and have always been popular. Poet Carl Sandburg had a well-known herd of Toggenburgs.

Toggs, as they're sometimes affectionately called, are always some shade of brown with a white or light stripe down each side of the face, white on either side of the tail on the rump, and white on the insides of the legs.

Toggenburgs rank slightly behind Alpines in milk production, with an average of 1,968 pounds of milk and 3.2 percent butterfat. But a Toggenburg currently holds the all-time record, with 7,965 pounds — an astounding 995 gallons of milk a year, from one little goat!

Toggenburgs have white markings on the face and rump.

It's in the Genes

In goat genetics, white is dominant and black is recessive. The white color pattern on Toggenburgs, including the vertical white stripes on both sides of the face, is dominant.

LaMancha

The LaMancha is a distinctly American breed. There's no mistaking a LaMancha: it looks like it has no ears!

During the 1930s Eula F. Frey of Oregon crossed some short-eared goats of unknown origin with her top line of Swiss and Nubian bucks. The result was the LaMancha.

If you show LaManchas at the county fair you'll have to put up with many exclamations of "What happened to the ears!" Some people who are somewhat more knowledgeable about livestock will accuse you of allowing the animals' ears to freeze off. But you don't milk the ears, LaMancha backers say. These goats have excellent dairy temperament, and they're very productive. A good average is 1,699 pounds of milk with 3.9 percent butterfat.

La Manchas are noted for their "lack" of ears and are claimed by some to be the most docile breed. They also tend to be good milk producers.

Dwarf Breeds

Two other breeds of goats are relatively new to these shores, and at least until recently, have been considered more of a novelty than true dairy animals. However, they have become firmly established as pets, and some people do milk them.

Nigerian Dwarf

The Nigerian Dwarf is of particular interest to backyarders. Introduced in the early 1980s when they were seen mostly in zoos, some of these little imports are excellent milkers for their size. As more serious breeders continue to develop them, their milk production is constantly increasing. What's more, they are considered dual-purpose animals, providing both milk and meat. The Nigerian Dwarf was the breed chosen for the Biosphere 2 experiment, in which eight people spent two years (1991–1993) sealed inside a self-contained, mostly self-sufficient dome along with 3,500 plant and animal species and no outside supplies or support except electricity. Biosphere 2 was designed as a space-colony model, though ecological research became the primary, scientific goal. At any rate, future space travelers might be milking Nigerian Dwarfs!

One Nigerian Dwarf doe gave a whopping 6.3 pounds of milk on test day, and another had 11.3 percent butterfat. A well-bred and well-managed Dwarf can be expected to produce an average of a quart a day over a 305-day lactation. Many of these good producers have teats as large as those of the full-sized breeds and are milked just as easily.

Nigerian Dwarf conformation is similar to that of the larger dairy breeds. Parts of the body are in balanced proportion (unlike the Pygmy, which looks like a beer keg with legs). The nose is straight, ears upright, and any color or combination is acceptable. Does are 17 to 19 inches tall; bucks, 19 to 20 inches tall. Weight should be about 75 pounds. Being oversized for the breed standard is a disqualification, as are a curly coat,

The Nigerian Dwarf is a dual-purpose animal and an excellent choice for the small farmer.

Roman nose, pendulous ears, and evidence of myotonia (a muscle condition characteristic of "fainting" goats.)

Nigerian Dwarfs offer several advantages to the home dairy. Three Dwarfs can be kept in the space needed by one standard goat, so with staggered breedings, a year-round milk supply is easier to achieve. This is enhanced by the Dwarf's propensity to breed year-round. (Compare this with seasonal breeding, discussed in chapter 10.) These small goats can be kept on places that might not have room for larger animals. Also, for some people, a regular goat will produce too much milk, while the Dwarf's quart or so daily is just fine. And the smaller animal is obviously easier to handle and transport. Many of these attributes appeal especially to elderly people.

One potential disadvantage: many people still regard Nigerian Dwarfs as pets. If you purchase one of these, chances are the goat has never been milked and the former owner has no idea how much milk the goat could produce. Animals like this, that have not been upgraded and bred for milk production, are not ideal choices for the home dairy.

Nigerian Dwarfs are recognized by the American Goat Society (AGS) and the International Dairy Goat Registry, but not by the American Dairy Goat Association (ADGA).

African Pygmy

Another dwarf breed gaining in popularity is the African Pygmy, often referred to as simply the Pygmy. This breed was first seen in the United States in the 1950s, and then only in zoos. These little goats are only 16 to 23 inches tall at the withers at maturity and does weigh only 55 pounds. They are very cobby, or stocky, compact, and well-muscled, quite unlike a standard dairy animal.

Despite their tiny size, some Pygmies are said to produce as much as 4 pounds of milk a day — that's half a gallon — and 600 to 700 pounds a year. And while the lactation period is

The African Pygmy can be productive but is not considered a dairy goat.

shorter than for full-size goats (4–6 months rather than 10 months), the butterfat content often exceeds 6 percent.

The Pygmy is more likely than the other breeds to have triplets or even quadruplets. They are registered by the National Pygmy Goat Association.

Other Breeds and Uses

Some new breeds are being created. (Mating a doe of one breed to a buck of another produces crossbreds. Creating a new breed is much more involved than that and generally takes years.) These include the *Kinder* (a Pygmy/Nubian cross), *Pygora* (Pygmy/Angora), *Santa Theresa* (another dual-purpose breed), and others. Although these have very enthusiastic, usually regional backers, they are rare compared with the six recognized breeds, and most are still in the early stages of development.

You might hear about a few other rare breeds, such as the Tennessee Fainting goat or Wooden Leg, which goes by several other names as well. These have the strange attribute of "fainting" when startled, such as by a loud hand clap. Formerly a curiosity, many have been developed as meat goats.

Raised primarily for their long silken mohair, Angora goats have become quite popular in recent years. They are also raised for meat. While the mohair aspects are beyond the scope of this book, basic feeding, breeding, and management are similar for Angora and dairy goats.

Angora goats are raised primarily for mohair and meat.

And then there are meat goats — animals raised for that specific purpose. The demand for goat meat has grown tremendously in recent years, due largely to ethnic markets (see chapter 15). Spanish and Angora goats are the traditional meat animals in the United States, but the Boer, originating in South Africa, has become hugely popular among meat goat ranchers. A male Boer fetched as much as $70,000 a few years ago, though prices are more reasonable today. But you don't have to pay for a special goat for meat for a family. I've never heard of anyone milking a Boer or other meat goat, but dairy goats provide plenty of meat as a by-product when culls and unwanted kids are butchered.

Goats have proved useful as working animals, too. *Wethers*, or neutered males, are commonly used for packing, and goats of any breed can be trained to pull a small cart or wagon.

Selecting a Breed

Which breed is best? There is no answer to that question. If your reason for raising goats is to have a home milk supply, a goat that produces 1,500 pounds of milk a year is as good as any other goat, regardless of breed, that produces a like amount. You might not need or want a purebred at all, at least at first. Mixed-breed goats are much easier to find, usually cheaper, and in some cases produce more milk than purebreds.

Even for those interested in purebred stock, the choice of a breed isn't made because of any breed superiority or rational factors. In most cases, a breeder just "likes the looks" of a particular breed. You'll also find it easier to find certain breeds than others, because the popularity of each varies from place to place. You might get a certain doe just because she's available, but if this means that stud service will be convenient and there is more likely to be a market for her kids, you'd be making a wise choice.

So, You Want a Goat?

This brief look at some of the basic facts about goats should help you decide if you really want to raise goats. I hope you do — but with full awareness of what will be expected of you. That means you'll want a lot more information on care and management. But before we get to that, let's take a closer look at the product that probably led you to goats in the first place: milk.

MILK

One of the first questions a prospective goat owner who is interested in a family milk supply asks is, "How much milk does a goat give?"

While the question is logical and valid, it's something like asking how many bushels of corn an acre of land can produce. How good the soil is, how much fertilizer is applied, what variety of seed is planted, how much of a problem weeds and insects are, and the amount of heat and moisture at the proper stages of development are all factors that affect the outcome.

To put this in terms that might be more readily understood by city dwellers, how many ladies' coats can a merchant sell? It depends on whether the seller is in downtown New York or on the edge of a small southern village; on whether the coats are mink or cloth; whether it's June or December; and so on.

How Much Milk?

There can be no set answer to the question of how much milk a goat will give, but here are some considerations.

Lactation Curves

It must first be understood that all mammals have lactation curves that, in the natural state, match the needs of their young. Humankind has altered these somewhat through selection to meet human needs, but they're still there.

The supply of milk normally rises quite rapidly after *parturition* (kidding, or freshening, or giving birth) in response to the demands of the rapidly growing young. In the goat, the peak is commonly reached about 2 months after kidding. From the peak, the lactation curve gradually slopes downward.

This brings up what is probably the most common problem with terminology in reference to production. We often hear of a "gallon milker." The term has little or no practical value, because we want to know at what point in the lactation curve this gallon-day occurred, and even more importantly, what the rest of the curve looks like. The goat that produces a gallon a day 2 months after kidding, then drops off drastically and dries up a short time later, will probably produce much less milk in a year than the animal whose peak day is less spectacular but who maintains a fairly high level over a long lactation. Especially in the home dairy, where a regular milk supply is the goal, slow and steady is more desirable than the flashy one-day wonder.

In addition, a "gallon" is neither an accurate nor a convenient unit of measure for milk. Milk foams, and what if a goat gives just over or under a gallon? A gallon and 1 cup is tough to measure and even tougher to record. It's much more practical to speak of *pounds of milk per lactation*. A gallon, for all practical purposes, weighs 8 pounds. The traditional lactation period is 305 days. If a goat is to be bred once a year and dried off for 2 months before kidding for rest and rehabilitation, this period is logical. The 305-day lactation period is simply an average; most goats milk for more or less than 305 days. According to the U.S. Department of Agriculture (USDA), only one-third of all does with official Dairy Herd Improvement Association (DHIA) records milked for 305 days. The milk production of many does declines sharply with the onset of seasonal *estrus,* or heat periods; after estrus, the does are dried off.

The average 305-day lactation period is a convenient way to compare animals (cows are judged in similar fashion), but it is mainly for record purposes. The backyard goat dairy has no need to adhere to such a schedule, and in practice even most commercial dairies milk an animal for shorter or longer periods depending on the animal's production. In some cases, it might not be worth dirtying the milk pail for a quart or so. In others, even a cup of milk might be considered valuable.

Actually, many household goat dairies with animals that exhibit long lactations would do well to milk them for as long as they can without rebreeding. Production could be lower the second year, but this would be offset by avoiding a 2-month layoff, breeding expenses, and unwanted kids — including the considerable amount of milk they will drink if not

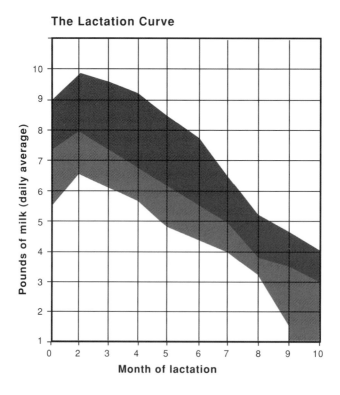

The Lactation Curve

The best doe in the herd produced 2,150 pounds of milk in 10 months (■). The lowest record shown is 1,300 pounds in 9 months (□). The average for the entire herd was 1,730 pounds in 10 months.

disposed of at birth. It should be pointed out, however, that not many goats will milk for that long: most will be dry before the 10 months are out. (See appendix A, Milking Through.)

Average Production Levels

Looking at averages can be meaningless — after all, how many American families *really* have 1.7 children? — but sometimes that's the only way to get even a rough idea of a situation. Just remember that when a breed averages 2,000 pounds of milk, some of the goats are producing 3,000 pounds, and some are only milking 1,000 pounds, and some even less than that.

For many years we said a decent average was 1,500 pounds a year. There are some indications that this average is increasing. (In 1998, for instance, the average of all six breeds was 1,794 pounds.) But then the question arises, is it increasing because more good goats are on test while less productive animals aren't? Fifteen hundred pounds — about 187 gallons — is still a reasonable expectation for the beginning goat farmer. But bear in mind that this average of 187 gallons over a period of 305 days doesn't mean you can plan on 0.6 gallons a day. Remember the lactation curve.

Breed Records

Breed records are even more meaningless for the home dairy than are averages. The new goat owner has about as much of a chance of even coming close to record production as the guitar pickin' kid down the road has of coming up with a hit song. It takes knowledge, experience, work, and maybe even a few lucky breaks, to produce a winner in any field.

At least the records will show you what goats are capable of. And they also demonstrate what a bucket of worms you get into when you ask, "How much milk does a goat give?"

You won't start out with a record setter, and you hope you won't get stuck with an under producer, but it would be nice to find one that's "average." The only way to know for certain how much milk a goat gives is to milk her, weigh the milk, and record it for the entire lactation period. Or purchase a goat from someone who's been doing that.

Sample Breed Records

The Nubian breed record of 5,940 pounds of milk was set in 1996, but the average for that year was only 1,567 pounds, or just a little more than one-fourth as much! And by definition, many were below the average.

The Saanen breed record was set in 1997 with 6,571 pounds of milk, but the 1998 average was only 1,899 pounds, or 4,672 pounds less.

The all-time all-breed U.S. production record was set by a Toggenburg in 1997: 7,965 pounds. (It might be of interest to note that in 1975, the record was 5,750 pounds, set by a Toggenburg in 1960. Today, four breeds have surpassed that mark.) The record-holder produced more than 3 gallons a day for 305 days.

At the opposite extreme, there are goats that freshen without enough milk to feed the barn cats.

Using Production Records

Leaving the pacesetters for a moment, let's look at the lactation curve on page 18 again. These are actual production records of a small herd of Nubians. The top doe produced 2,150 pounds in 10 months, the bottom doe 1,300 in 9 months. Notice the lactation curve. The average production goes from 7½ pounds at kidding to about 8 pounds 2 months later. From there it tapers off to about 3 pounds at 10 months after kidding.

The lactation curve on page 21 provides another example of an "average" small herd. These are actual, individual records from a herd of four grade does. It shows how much production can vary among animals. One doe had a 17-month lactation. She gave 1,800 pounds in the first 10 months and continued to produce a steady 5½ pounds daily until pregnancy caused production to drop. Another doe reached her peak at 4 months.

If you owned these four does and were going to sell one, which one would it be? There are two lessons here:

First, remember this when you *buy* a goat: are you buying an animal someone is culling because of low production? Ask to see milk records.

Second, without records and perhaps a chart like this one, no matter how rough, you don't know for sure what's happening. Not a month from now, not a year from now, and certainly not 5 years from now when you're

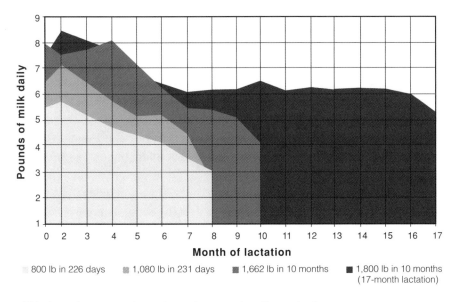

800 lb in 226 days   1,080 lb in 231 days   1,662 lb in 10 months   1,800 lb in 10 months (17-month lactation)

This lactation curve chart shows how much milk production can vary among goats. The goal of the home dairy is to start out with the best milkers available, and then to improve the herd through breeding and selection.

trying to decide which granddaughters of your present milking does to keep and which ones to sell or butcher.

Note that one of these does produced more than twice as much as another, even though they ate about the same amount of feed and required the same amount of care. Note also that two weren't worth milking after only 8 months. And finally, it should be obvious that when you milk your own goats, you don't have as steady a milk supply as when you pick up a gallon from the grocery store whenever you need it. There will be times when you could drown in milk, and other times when you'll eat your cornflakes dry.

Why so much variation? Age is a factor in milk production. Records from this same herd show that peak production comes in the fourth or fifth year. But there are also other factors involved, and any individual goat can vary erratically from one year to the next. Even on a day-to-day basis, milk production is affected by changes in weather, feed, sickness or injury, outside disturbances, and other factors.

The question "How much milk does a goat give?" is best answered by another question: "How long is a piece of string?"

Discovering a New Taste

From all this you can assume that you'll be able to find a goat that will produce a respectable amount of milk for your table for at least a part of the year. Next question: will your family drink it?

Many goat raisers delight in telling stories about how they tricked finicky people into drinking goat milk and how those people couldn't tell the difference. In taste tests, most people will actually prefer goat milk to cow milk.

However, sometimes people taste the milk, find it delicious, discover it's goat milk, and suddenly decide that it's really awful. There isn't much hope for people like this, aside from perhaps serving goat milk from a bottle or carton that came from a cow dairy and not telling them the truth. (I'm not advocating this, but some people have successfully used this technique on their children.)

Goat Milk as "Medicine"

When you've been raising goats awhile you'll surely be asked what you do with the milk. Many people seem to assume that goat milk is used only in hospitals. The healthful aspect of goat milk is as legendary as goats' aroma and their preference for tin cans.

Many doctors do, in fact, prescribe goat milk for certain conditions. Many more would, if steady supplies of sanitary goat milk were available. Goat milk might be recommended in cases of dyspepsia, peptic ulcer, and pyloric stenosis. It is preferable to cow milk in many cases of liver dysfunction and jaundice because the fat globules are smaller (2 microns versus 2½ to 3½ microns for cow milk). These smaller fat globules provide a better dispersion of fat in the milk and, thus, a naturally homoge-

---

Did You Know?

When whipped, cream from goat milk is bulkier than cow cream; the specific gravity of goat milk is 0.83 and that of cow milk 0.96.

---

nized mixture. (There is more involved in the "natural homogenization" of goat milk than the size of the fat globules, as we shall see shortly.)

Goat milk has been used for infants during weaning, infants with eczema, children with a liability to fat intolerance or acidosis, pregnant women troubled by vomiting or dyspepsia, and anxious or elderly people with dyspepsia and insomnia.

Goat milk is more easily digested by humans than is cow milk because the fat is finer and more easily assimilated. It is particularly rich in antibodies, and when freshly drawn has a much lower bacterial count than cow milk.

Exploding More Goat Milk Myths

Despite its numerous health benefits, goat milk is *not* medicine. It's food. Good food! More people in the world consume goat milk, either as a drink or made into cheese or yogurt, than they do cow milk. The United Nation's Food and Agriculture Organization estimates that worldwide, humans use 4.8 million tons of goat milk per year.

Because most Americans aren't familiar with goat milk, they continue to have misconceptions about it — misconceptions that seem to be based on the comic-strip image of the goat, or on the unfortunate experiences of a few who have been exposed to goat milk produced under conditions that make it unfit for human consumption. Actually, Americans have educated their palates a bit and are now at least a little more open to the wonderful flavors of the many kinds of goat cheese that are available. But the milk still suffers from bad press.

The home winemaker who takes a basket of overripe and spoiled, wormy, moldy fruit, puts it in a dirty crock, and pays no attention to proper fermentation, does not end up with a fine wine. And the goat raiser who milks a sickly, undernourished animal of questionable breeding into a dirty pail, lets the milk "cool" in the shade, and serves it in a filthy glass does not end up with fine milk. While few people would disparage all wines after tasting the concoction just mentioned, many people are all too willing to write off all goat milk after one unfortunate experience.

Milk is as delicate a product as fine wine. It must be handled with knowledge, and care, whether it comes from a cow or a goat or a camel.

Tastes Funny? Specifying Off-Flavors

Sooner or later we encounter milk that "tastes funny." Although some are quick to attribute the flavor to the goat, this usually isn't the case. But we need a better description than "funny" or "off-flavored" to pinpoint the cause.

Trained milk tasters describe off-flavors in ways that can help reveal the cause. A *bitter* taste can result from the doe eating weeds or strong feeds shortly before milking, or by conditions present in the milk in late lactation. *Coarse* or *high-acid* milk can indicate bacterial growth. *Barny* milk is self-descriptive and can be caused by milking in a smelly barn, or not moving the milk from the barn quickly enough, or if the doe has been breathing the air in a poorly ventilated barn, even if she's milked in a clean parlor.

A *disinfectant* flavor indicates a residue of chlorine or other disinfectants in poorly rinsed equipment. *Feed* flavors can be caused by wild onions and garlic, turnips, silage, and even green grass. Eliminate access to such feeds for 2 to 4 hours before milking. *Foreign* is used to describe flavors resulting from inhaled odors such as fresh paint, petroleum products, and fly or other sprays. A *malty* flavor usually means the milk wasn't properly cooled. *Metallic* flavors are caused by milk coming into contact with corrodible metal, including copper and rust. *Musty* flavors can come from moldy hay or stagnant water. *Salty* is most common in late lactation, and often indicates mastitis. A *rancid* flavor should not be a problem with fresh milk, but a form of it can be caused by the enzyme lipase, which causes a change in the composition of milk fat. This can usually be traced to extreme agitation of warm raw milk in air, which can occur when too-vigorous milking results in foaming or when a glass container has been left in the sun for even a short time. *Utensils* suggests an "unclean" flavor attributed to inadequate washing and sanitizing of equipment that comes into contact with the milk, but can also be caused by the animals drinking dirty water.

What about reports of "odd" or unpleasant tastes in goat milk? One of the more common ones, poor sanitation, is also the most easily remedied. Yes, it can happen, and there are many possible causes. Strict sanitation means more than just "clean enough to eat off of." It includes thorough washing of equipment in hot water with a stiff brush, not a washcloth; rinsing with a chemical sanitizer or scalding hot water; avoiding pitted or scratched aluminum containers; milking in a clean well-ventilated place free of odors; and filtering the milk through an appropriate material (a milk filter made for the purpose, not reused cheesecloth).

Other off-flavors can be eliminated by rapid chilling; keeping milk out of sunlight or fluorescent light; not exposing milk to copper or iron; not mixing warm raw milk with cold or pasteurized milk; and avoiding violent agitation.

Nutritional deficiencies, such as lack of cobalt, vitamin B_{12}, and vitamin E, may also contribute to an off-taste. Some off-flavors can be traced to unsaturated fatty acids in milk that are more susceptible to oxidation, hence causing off-flavors. This can be corrected by decreasing the amount of fat in feed (reduce amount of soybeans and whole cottonseed, for example) and by increasing the forage-to-concentrate ratio.

Note that in some places in the world, a "goaty" flavor in cheese is highly prized. In Scandinavia, for example, milkers are selected and bred for this trait.

In rare cases, animals can give off-flavored milk. We can't call it "goaty" because some cows have the same problem. Off-flavored milk can be caused by strong-flavored feeds and plants such as ragweed, grape leaves, wild onion, elderberry, honeysuckle, and many others. This is one reason many commercial dairies don't let their goats browse or graze: after all, if goats are on pasture, it's impossible to control what the animals eat.

Off-flavors can also be caused by mastitis, which often makes milk taste salty. A rancid flavor can result when a doe is in late lactation or when foamy milk is cooled too slowly. (The temperature should drop to under 45°F in less than an hour.) A "cardboard" flavor might be caused by oxidation resulting when milk is left in the light or when there is copper or iron in the milk container. Obviously, a dirty barn, dirty goats, and generally unsanitary practices make milk an excellent environment for bacteria, which can not only affect the taste but produce illness.

Goat Milk vs. Cow Milk

Goat milk does not taste any different than cow milk. It doesn't look appreciably different. It is somewhat whiter, because it doesn't contain the carotene that gives a yellow tinge to the fat in cow milk. Goats convert all carotenes into vitamin A. It is not richer. It certainly does not smell. If it does, something's wrong.

Most goat raisers enjoy serving products of their home dairy to skeptical friends and neighbors. The reaction is invariably, "Why, it tastes just like cow milk!"

I have noticed, however, that city people who are accustomed to regular standardized milk — milk that has butterfat removed to just barely meet minimum requirements — or worse yet, who drink skim milk, are prone to comment on the "richness" of goat milk. They'd say the same thing about real cow milk if they had the opportunity to taste it before the technologists messed around with it and turned it into chalk water. As with cow milk, the percentage of butterfat — the source of the "richness" — varies with breed, stage of lactation, feed, and age. But generally, there is virtually no difference in taste or richness between whole, fresh cow milk and goat milk.

Biochemical Composition

Despite these similarities, the *composition* and *structure* of the fat in cow milk and goat milk is one of their more significant differences.

It has long been said that goat milk is "naturally homogenized" because of its small fat globules. Actually, it turns out that it probably isn't the size of the fat globule that causes the cream in goat milk to remain in suspension. Recent research has shown that goat milk lacks a fat-agglutinating protein, an *euglobulin*, that would cause the fat globules to adhere to one another and mass up. In fact, the cow is probably the only domestic animal that produces milk with this particular protein, according to Professor Robert Jenness of the University of Minnesota. Sow and buffalo milk do not form creamlines. Because homogenized cow milk is the norm in the United States, most people would probably be surprised to see cream rise in milk. They won't be shocked by goat milk.

Still, from the standpoint of human health, natural homogenization is better. Some research has shown that when fat globules are

Average Composition of Goat and Cow Milk*

ELEMENT	GOAT	COW
Water (%)	87.5	87.2
Food energy (kcal)	67.0	66.0
Protein (g)	3.3	3.3
Fats (g)	4.0	3.7
Carbohydrate (g)	4.6	4.7
Calcium (mg)	129.0	117.0
Phosphorus (mg)	106.0	151.0
Iron (mg)	0.05	0.05
Vitamin A (IU)	185.0	138.0
Thiamin (mg)	0.04	0.03
Riboflavin (mg)	0.14	0.17
Niacin (mg)	0.30	0.08
Vitamin B$_{12}$ (µg)	0.07	0.36

*Composition per 100 g.

Note: Charts like this can be confusing. There are many similar charts, but seldom do two agree. The reason is that the composition of milk varies with the animal, the stage of lactation, feed, and other factors. Averages of one study can differ from averages of another, and neither "average" is likely to apply to your goats.

forcibly broken by mechanical means, an enzyme associated with milk fat (*xanthine oxidase*) is freed. This enzyme can penetrate the intestinal wall, enter the bloodstream, and damage the heart and arteries, creating scar tissue. In response, the body may release cholesterol in an attempt to lay a protective fatty material on the damaged and scarred areas, which can lead to arteriosclerosis. According to Dr. G. F. W. Haenlein, Cooperative Extension Dairy Specialist at the University of Delaware, no such problems are associated with natural (unhomogenized) cow milk or with goat milk.

Goat milk also has a higher amount of shorter-chain fatty acids than does cow milk. Glycerol ethers, too, are much higher in goat milk (which is important for the nutrition of the nursing newborn), and there is less orotic acid, which can be significant in the prevention of fatty liver syndrome.

The protein content of goat and cow milk is fairly similar. Goat milk has more vitamin A, more B_1 and B_3 vitamins (thiamine and niacin), but less B_6 and B_{12}. Cow milk is marginally higher in lactose.

As for minerals, goat milk is higher in calcium, potassium, magnesium, phosphorus, chlorine, and manganese, while cow milk is higher in sodium, iron, sulfur, zinc, and molybdenum. Goat milk has smaller amounts of certain enzymes, including ribonuclease, alkaline phosphatase, lipase, and the xanthine oxidase mentioned earlier.

While there are some differences between goat milk and cow milk, they're minor and their nutritional significance to humans, if any, hasn't been documented.

Raw Milk vs. Pasteurized

If you want to start a dandy (and sometimes heated) discussion in goat circles, just casually bring up the topic of raw milk and stand back.

For obvious reasons, most people never worry about such trifles. They just buy their jug or bag of milk and don't even have to think about where it came from or how it was treated. For those of us interested in simple living, however, or even just in producing our own dairy products, it's not nearly so simple.

Here's the problem. Milk is the "ideal" food, for animals, for humans — *and* for bacteria. Milk is extremely delicate. It can attract, incubate, and pass on all sorts of nasty things such as salmonella, toxoplasmosis, Q fever, listeriosis, campylobacteriosis, and others that most of us non-medical people never even hear about. In short, nature's most healthful food can make you very sick.

The Industrial Age answer has been pasteurization — heat-treating the milk to kill or retard most of those threatening organisms. Government regulations now demand that such treatment be performed "for the public good."

This is no doubt a wise policy, for the masses. When you pick up a jug of milk at the store, how would you know if the dairy animals were healthy, if the milker had clean hands or a runny nose, and if proper sanitation measures were taken, without such governmental intervention?

Pasteurization 101

If you decide to pasteurize your milk, all it involves is heating it to 165°F for 15 seconds. Home-size pasteurizers are available. (Do *not* use a microwave, as some people suggest. Dairy scientists have proved that it doesn't work.) And if you prefer raw milk but want an extra measure of safety, have your goats tested for tuberculosis, brucellosis, and campylobacter. And be sure to practice scrupulous sanitation. (For more on specific sanitation procedures, see chapter 13.)

But many goat-milk drinkers raise their glasses with a different perspective. They know everything about their animals, from age and health status to medical history to what they ate since the last milking. The home milker knows exactly how the milk was handled — how clean the milking area and utensils were, how quickly the milk was cooled and to what temperature, and how long it has been stored. (Usually, it hasn't been stored long. Most goat milk from the home dairy is probably consumed before cow milk even leaves the farm, if it's only picked up every 2–3 days.)

Under these conditions, many people who milk goats feel it isn't necessary to go to the bother of pasteurizing their milk. Also, many people raise goats because they *want* raw milk. (It's illegal to sell raw milk, goat or cow, in most states, although goat milk is often sold as "pet food," for orphaned or sickly young animals.) Some think it tastes better. And some say it's more nutritious.

Pasteurization does have an effect on nutritive value. But raw-milk opponents claim it's very minor, and insignificant when compared to the potential dangers of untreated milk. To them those dangers are horrific: they would just as soon drink poison.

One of the potential problems raw-milk opponents often point to is campylobacteriosis, a gastrointestinal disease caused by campylobacter — a bacteria universally present in birds, including domestic poultry. The symptoms, which range from mild to severe, include abdominal cramps, diarrhea, and fever. Apparently, farm families who drink raw milk regularly build up an immunity. Most of the reported cases have involved farm visitors, those unused to raw milk. Raw-milk advocates point out that the incidence is very small — might as well worry about being struck by a meteorite, they say.

More serious diseases associated with raw-milk consumption are tuberculosis and undulant, or Malta, fever (called brucellosis in animals). While cattle are susceptible to tuberculosis, goats are highly resistant: they have not been implicated in tuberculosis outbreaks. And while undulant fever is a goat problem in some countries, including Mexico, it hasn't been in the United States.

So, who's right, those who oppose raw milk or those who advocate it? Probably both, in certain situations and under certain circumstances. For example, people with impaired immune systems, such as infants and the elderly, are more at risk for some of the minor diseases that can be passed from goats to humans. In general, my impression is that most people who feel strongly about this, one way or the other, base their decisions more on emotions than on facts. Your personal decision will very likely depend on your psychological makeup: how you regard science and medicine in general, for example, or your attitude toward natural or organic foods, or whether or not you fasten your seatbelt.

GETTING YOUR GOAT

So you've decided to buy a goat. Now the problem is finding one, and perhaps more importantly, finding the one that's right for you.

The popularity of dairy goats has soared in recent years. The U.S. Department of Agriculture (USDA) estimates that there are now about one million dairy goats in the country, an increase of about 20 percent in 10 years. (Compare that to 10 million cows). But goats are by no means common, and they certainly aren't distributed evenly around the country. There might be many dairy goats in your area, in which case finding the right one will be fairly easy. Or there might be none at all, which means you'll have to do some traveling to get your home dairy started.

Beginning Your Search

There are many ways to begin your search. If you have friends or neighbors who have goats, or if you have seen goats in your area, you have a good start. Almost everyone with goats has animals for sale sooner or later, and if they don't, they'll probably know of someone who does. They might even know of a goat club in your area, and then you'll have it made.

No doubt you've been watching the classified ads ever since you started thinking about getting a goat. Classified ads usually appear for only a few days, so you'll have to read them regularly and probably over a long period.

Check out the farm papers serving your area. Again, it might be a lengthy process if there are few goats in your region, but having a goat is worth it! Of course, you can also place your own "goats wanted" ad.

Be sure to attend county and state fairs and goat shows. This is often an opportunity not only to see and compare several breeds at once but also to have a good time and to talk to goat people. Reach them at home and they might have a dozen other things to do, but talking about goats is one reason they're at the show! And they might have goats for sale.

In most livestock-producing rural areas, you'll come across "sale barns." Again, the numbers of goats they handle will depend somewhat on how many goats are in the neighborhood, but even if they only see a few every couple of months, they might let you know when they have one or more if you tell them you're interested.

Goat shows and fairs provide good opportunities to see several breeds of goats at once, to learn more about goats by watching the judging, and to talk with goat people.

Sale-Barn Warning

Be especially cautious with sale-barn animals, however. Many of these are disasters, and even healthy-looking animals can pick up diseases in a sale-barn environment. Some good animals do pass through them, though, so they're worth a look.

Other local sources to contact for leads on goats include vocational-agricultural teachers, Future Farmers of America (FFA) and 4-H leaders, veterinarians, feed stores (if they sell goat feed, the managers will know who buys it), and county extension agents.

Of course, you'll also want to watch the ads in the national goat and livestock publications, such as *Countryside & Small Stock Journal* and *Dairy Goat Journal*. If you're lucky, you might learn of a knowledgeable and reliable breeder near you. Also, increasingly, the Internet can provide you with information on goats for sale.

These suggestions assume you'll want to buy your first goat close to home. When you've gained some experience, you might want a certain breed or bloodline and be willing to buy an animal hundreds or even thousands of miles away, maybe without even seeing it. But at first, it's much wiser and easier to do your searching closer to home. There are several good reasons for that. If you don't know very much about goats, you'll want to see some, and certainly the one you'll be spending a lot of time with. You'll probably want to talk to someone with experience, and study their setup. You'll almost certainly want to make some nearby arrangements for breeding. If you live in a goat-deficient region, you might not have much of a choice of breed or animals by shopping close to home, but unless you have plenty of time and money, that might be more acceptable than traipsing around the countryside.

Terms to Learn

If you're like most people who just want a family milker, you'll end up buying what's available regardless of breed, type, conformation, and even desirable traits. Many people start out like this. It's a good way to gain experience. But you'll still want to have some idea of what to look for, and it will be helpful to at least be familiar with some of the jargon the seller might toss your way. (On the other hand, by the time you finish reading this book, you might know *more* than the seller. I still chuckle when I recall the lady who showed me her Toggenburgs — which she called "Toboggan Birds." And she was serious!)

You're already familiar with the breeds of goats. But you'll probably encounter other terms that help identify and classify individual goats, such as registered purebred, AR (or Advanced Registry), star milker, grade, recorded grade, American, and linear appraisal. What do all these strange words mean, and how can you use them to help select the right goat?

Registered Purebred

A *registered purebred* animal is one with a pedigree that can be traced, through a registry association's herdbook for the breed, to the very beginnings of the recognition of the breed as "pure."

Some purebreds are not registered, for a variety of reasons. The breeder might not think it's good enough to warrant registering. In many cases smaller breeders who aren't really interested in registered animals simply don't want to bother with the paperwork and expense. In some cases, with a lot of work or luck or both, you can trace a purebred but unregistered animal back to its registered ancestors, and with the proper paperwork, have it registered. More commonly, it's impossible to prove that the goat truly is a purebred if it isn't already registered.

Pedigree and Registration Papers

A *pedigree* is merely a paper showing the ancestry of an individual animal — a family tree. *Registration papers* are official documents showing that the animal is entered in the herd book of a registry association.

AR

Another term you'll hear is AR, or Advanced Registry. An *AR* doe is one that has given a certain amount of milk in a year. The amount varies with the age of the goat and other factors, but the AR designates the animal as a good milker.

Star Milker

Still another term referring to production is *star milker*, or ★ milker. Unlike the AR, the star is based on a 1-day test rather than on an entire lactation. Like the AR, points that count toward stars are based on the stage of lactation the doe is in. It should be noted that many does who have earned AR certificates cannot become ★ milkers because they never give enough milk in 1 day. Even though weighted for the length of lactation, a doe still has to produce about 10 to 11 pounds of milk in 1 day to earn a star. Conversely, many ★ milkers can't earn their AR because they don't produce enough in the entire 305-day lactation to qualify. Both are official ratings, however, and both are supervised by someone other than the owner.

Calculating ★ Milkers

The star is granted for a minimum of 18 points. Points are based on:

- ◆ **Total pounds of milk produced in 24 hours** (points equal the sum of pounds of milk given).

- ◆ **Total number of days in the current lactation,** with 0.1 percent for each 10 days with a 3-point maximum (multiply number of days by 0.01).

- ◆ **Butterfat percentage** (multiply the percent by the points/pounds given per day, then divide by 0.05). For example, a doe that gives 6.4 pounds in the morning and 7.0 pounds in the evening would earn 13.4 points. If the doe was fresh for 44 days, 44 is multiplied by 0.01, adding 0.44 points for lactation. If the butterfat is 3.7 percent, 0.037 times 13.4 equals 0.4958 pounds of fat. This amount is divided by 0.05, yielding 9.91 points. The doe's total points: 23.75. She gets a star!

If a doe's dam has earned a star and she earns one herself, she becomes a ★★ milker. If her granddam also had a star, she is a ★★★ milker. Bucks can also have stars, signifying those earned by their maternal ancestors.

Grades and Americans

An animal without a pedigree is considered a *grade*. It could be pure-bred, but without the papers, you can't prove it. Most grades are mixtures of two or more breeds. Some grades are very good animals. If such a goat meets certain requirements, it might be listed as a *recorded grade* of whichever breed it most closely resembles.

If such a recorded grade doe is bred to a registered buck, the offspring will be one-half "pure." The does from these matings can be recorded grades. If one of these does is bred to a purebred buck, the kids will be three-fourths purebred. One more generation will result in a goat that is seven-eighths "pure," and one more will produce kids that are fifteen-sixteenths pure.

Seven-eighths does and fifteen-sixteenths bucks can be registered as *Americans*. You might also encounter *NOAs* (Native on Appearance), and *Experimentals* (which usually means a buck jumped the fence and bred a doe of a different breed).

Which Goat Is Best for You?

Assuming you have a choice of goats from all these classifications, which is best? Once again, there is no simple answer. For a home dairy, a regis-tered goat, or even an unregistered purebred, may or may not be the best choice. If you're keen on showing, you'll want registered purebreds. But our emphasis here is on milk.

Registration papers don't mean anything more than that the goat is listed with one of the registry associations and that the pedigree can be traced back to the closing of the herd book. This is important to experi-enced breeders who are trying to upgrade and improve their animals. But if all you want is milk, and you know nothing about goats' family trees, the papers don't mean much. In fact, some registered purebreds are very poor milkers. A registration certificate is not a license to milk.

On the other hand, many grades, and even brush goats or scrubs, turn out to be excellent milkers. One illustration of this comes from a lady in a southern state who bought a brush goat. It appeared to be of good Toggenburg breeding, but it had been running semiwild, as brush goats do, clearing a patch of hilly land so cattle could graze there. These goats are not fed hay or grain, and they aren't milked. When this particular doe was treated as a dairy goat, she turned out to be a superb milker! She had always had the genetic ability but had fallen into the hands of someone who didn't know or care. With proper feed and management, she blossomed. There are many "common" goats like this.

In other words, you might find very fancy, nice-looking, papered goats that won't produce enough milk to feed the cat, and you might find rather ordinary, crossbred, inexpensive goats that will fill your gallon milk pail to the brim. Your problem as a prospective buyer is, how do you know if a particular goat will be a reliable and efficient milker? This is a two-prong question. We want to know if a goat will produce milk, and we want to know if she will be efficient, which is to say, economical.

Registration papers and pedigrees might tell you something about her milking ability, but only if you know how to read them. Show wins (of the animal herself or of her ancestors) may or may not mean she's a good milker. You might look at stars, AR certificates, classification scores, linear appraisal scores, and more; but a goat that has none of these isn't necessarily a poor milker. It might just mean that the owner hasn't bothered to go after them.

Consider the Price

The upshot is that records and papers give you, not a guarantee, but some degree of insurance. And you pay for the insurance. Only you can decide if it's worth it, for any particular animal. If you pay hundreds, even thousands of dollars for a good goat, your milk will be that much more expensive.

But it's even more complex than that. As we'll see in chapter 14, you might lower your milk costs by selling registered purebred kids,if you can sell them for a good price. On the other hand, as a beginner, you might be more comfortable "learning" with a less expensive and perhaps more adaptable animal. To all this add the fact that sometimes there are very good purebreds for sale at very reasonable prices, and rather poor grades for sale at exorbitant prices.

Consider the Source

How do you cut through this maze of confusion and conflict? Very much like you buy a car, or anything else. You arm yourself with as many facts as possible; you ask some questions; you rely on the integrity and reputation (and probably even the personality) of the seller and the seller's place of business; and then you buy a particular model just because you like the color or because it "appeals" to you!

A Buyer's Checklist

Here are some questions to ask as you begin to narrow your choices:

- **Do you have faith and confidence in the breeder?** Does the breeder appear to be knowledgeable about goats? Is he trustworthy? You're going to have to rely on that person's experience, honesty, and integrity, to a large degree, for your first purchase.

- **What are the goats' living conditions like?** Are they running loose, or are they staked out in a weed patch? Are they well housed, in a neat and comfortable building? Do they have an exercise yard with good fencing?

- **How do the animals look?** Are they disbudded? Neatly trimmed? Do their hooves show obvious care?

- **Are there papers or other official records you can see?** If all the owner has to show you are "barn records," you'll have to rely on the owner's honesty. If there are no records at all, it could mean that the seller doesn't know very much about goats or doesn't care very much about them.

Spotting a "Good" Goat

In all seriousness, it will be helpful for you to have some idea of the differences between a "good" goat and a "poor" one before you purchase an animal. Making this distinction takes a great deal of experience, but following are some general tips to help you get started.

Conformation

The general appearance of livestock, the way an animal is put together, is called *conformation*. Conformation is what a dairy goat judge is looking at when placing animals in the showring. While a licensed judge spends many hours and years of study and practice to learn his trade, to a certain degree you must make yourself a "judge" when you buy a goat and as you build and improve your herd.

To become an informed buyer, you must learn the parts of the animal. You must know what good animals look like, and what traits are considered defects. You must be observant enough to see both good conformation and defects, and knowledgeable enough to weigh and evaluate their relative importance. Some people have a sixth sense for evaluating livestock, and others never get the hang of it. While you'd have to see and handle hundreds of animals and study far beyond the scope of this book to become even a fair judge of goats, familiarizing yourself with some key concepts will make a big difference when you purchase your first doe.

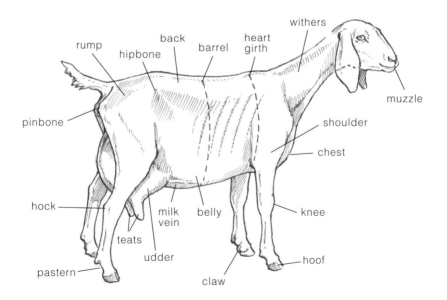

It's important to know the body parts of a goat when discussing the animals with judges, veterinarians, and other breeders.

Shape and Carriage

First, observe the animal from a distance. Note the doe's general shape and carriage. She should be feminine, with a harmonious blending of parts. The show scorecard speaks of "impressive style, attractive carriage, and graceful walk," but this is no beauty contest! These traits can tell a great deal about her general condition, vigor, and the dairy character that means milk in the pail.

Then move in for a closer look.

Head

The head should be moderately long with a concave or straight bridge to the nose, except in the Nubian, which must have a definite Roman nose. Saanens have a concave nose, or dish face. The eyes should be bright, the forehead between the eyes broad.

Ears

Ears are a part of conformation, but they're of small importance if your primary concern is having fresh milk. As LaMancha breeders say, you don't milk the ears. But to make this description more complete, note that the ears should be pointing forward and carried above the horizontal, again with the exception of the Nubian, which must have a long, thin-skinned ear, hanging down and lying flat to the head. LaManchas, the "earless" breed, have a size limit of about 1 inch per ear. The so-called airplane ears that result from a cross between a Nubian and another breed are ridiculed by many, but again, you don't milk the ears and more than a few people have such animals and love them.

Muzzle

Of more importance to the home milk supply are such points of conformation as the muzzle, which must be broad with muscular lips and strong jaws, as this is an indication of feeding ability. Large, well-distended nostrils are essential for proper breathing.

Neck

The neck should be clean-cut and feminine in the doe, masculine in the buck, with a length appropriate to the size of the animal. It must blend into the shoulders smoothly and join at the withers with no "ewe neck." The goat needs a large, well-developed windpipe.

Forelegs
The forelegs must be set squarely to support the body and well apart to give room to the chest.

Rib Cage, Chest, and Back
The rib cage should be well sprung out from the spine with wide spacing between each rib. The chest should be broad and deep, indicating a strong respiratory system. The back should not drop behind the shoulders, but it should be nearly straight with just a slight rise in front of the hipbones.

Hipbones
The hipbones should be slightly higher than the shoulder. The distance between the hipbones and the pinbones should be great, but not so long as to make the animal look out of proportion. The slope of the rump should be slight, and the rump should be broad. The broader the rump, the stronger the likelihood that the goat will have a high, well-attached udder — a desirable trait.

Barrel
The barrel should be large in depth, length, and breadth. A large barrel indicates a large, well-developed rumen necessary for top production.

Udders and Teats
There are many types of udders and teats. Avoid abnormalities such as double teats, spur teats, or teats with double orifices.

While very large, so-called sausage teats are undesirable, very small ones may be worse as they make milking difficult, especially for people with large hands. However, many first fresheners have tiny teats that quickly become more "normal" with milking. It's often easiest to let the kids nurse does like these.

Don't be too impressed by large udders. Many of them are just meat. For a pendulous udder, you'll have to milk into a pie pan because there isn't room to get a pail under the goat! Of more serious concern, pendulous udders are more prone to injury and mastitis infections.

A well-attached capacious udder, carried high out of harm's way, with average-size teats and free from lumps and other deformities, is the heart of your home dairy.

Skin

The condition of the skin reflects the general condition of the entire animal. It should be thin and soft, and loose over the barrel and around the ribs. A goat with unhealthy-looking skin probably *isn't* healthy; it might have internal or external parasites. Check for lice and mites.

If you're looking at younger stock, avoid the overdeveloped kid. A kid that develops too early seldom ends up being as good an animal as one that has long clean lines and enough curves to indicate that the framework will be filled out at the appropriate time.

Wattles

Many goats have *wattles*, which are small appendages of skin usually found on the neck, although they can be just about anywhere on the body. They are a family trait, not a breed characteristic. Some animals of all breeds have them; others don't. They are merely ornaments. Some breeders cut wattles off of young kids, not only to make the animal look smoother, but because sometimes another kid will suck on the wattles, which causes soreness. When wattles are removed with sharp scissors early in a kid's life, there is little bleeding. Wattles also can be removed by tying a thread tightly around the base, which causes the wattles to fall off.

Horns

Horns, likewise, are indicative of neither sex nor breed. Some goats have them, some don't. And many goats born with horns have them removed soon after birth because horns can cause many problems later. Disbudding young kids is much easier and safer than dehorning older goats. See chapter 7 for more details.

These ornaments are a disqualification for dairy goats that are shown.

The Whole Picture

Observe the goat from the side, the front, and the rear, with a critical eye. A good dairy animal has a classic wedge shape when viewed from above: she has a delicate neck rather than a bull neck, and the barrel is wider than the shoulders. The top line is straight; a severely sloping rump is a defect. In official judging, general appearance and breed characteristics are allotted 30 points; dairy character, 20; body capacity, 20; and the mammary system, 30; for a total of 100 possible points.

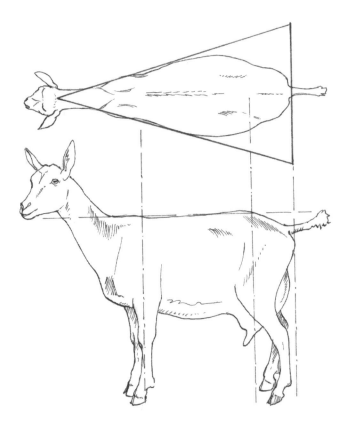

A good dairy animal has a "wedge" shape when viewed from above. A straight top line and moderately sloping rump are also indicators of good conformation.

Judge the Seller

If for you a "good" goat is simply one that milks well, first consider health, conformation, and overall appearance, because a sickly animal or one that isn't built like a dairy animal isn't going to do the job for you.

Next, consider any records that might be available. The easiest way for the beginner to judge papers and records, including barn records, is to evaluate the character of the seller. It might be more difficult to judge the owner than the goat, but at least you have a lifetime of experience with people! Without records of any kind, your best assurance of getting the goat that's right for you is by buying from someone you feel you can trust.

Such a person will help you learn, and some will even stand behind the animals they sell. (Maintaining such a policy sometimes asks a lot, as when careless or ignorant people take home a good goat, neglect or abuse it, and then complain that they were ripped off because the animal doesn't meet their expectations.) You want a seller you can call on later for help and advice. When you deal with someone like this, you get much more than just a goat for your money.

On the other hand, there are people who have been raising goats for years and who still don't know as much as you will after reading this book. While some people have 20 years' experience, these folks have 1 year's experience twenty times. There are also out-and-out crooks, who will sell worthless animals and overpriced animals. Some will try to sell you "registered" goats with the papers to come later (they never do), and there are others who are more interested in disposing of a goat or acquiring your cash than they are in helping you or promoting goats.

People who raise goats, in other words, are a cross-section of people in general. If you buy a goat without knowing very much about goats, it will help to know something about people.

There's a reverse side to this, too. Some buyers are pushy, obnoxious, know-it-alls. Not you of course, but bearing in mind that an experienced seller has most likely dealt with people like this in the past, be aware that you are being evaluated, too, and act accordingly.

Official Records and Barn Records

Most people think official milking records such as DHIA (Dairy Herd Improvement Association) are best, but even the owner's own barn records of daily milk production are better than none. Remember that the amount of milk produced in a year is more important than the amount produced on any one day. You don't expect your first goat to break any world records, but you want her to be more than a nonmilking pet, too!

Show wins can be impressive, and they'll tell you something about what qualified dairy goat judges think of the animal's conformation, but like registration papers, blue ribbons are no license to fill a milk pail. Again, registration papers and pedigrees mean little unless you're familiar with a great many names and backgrounds, and that won't come until later, after study and experience.

Be wary of milk records expressed in pints and quarts (and downright skeptical of milk recorded in gallons!). Even an honest and well-meaning milker can be misled by a bucket of foaming milk that "looks" like 3½ pints. Weight is much more reliable.

Official milking records will be in pounds and tenths of pounds, and so should unofficial barn records. For all practical purposes a quart of milk weighs 2 pounds; a gallon is roughly 8 pounds.

Barn records depend entirely on the accuracy of the scales and the integrity of the milker. They can be falsified or altered. Official tests, monitored by outsiders, are much more reliable, but because of cost and other factors, they aren't widely used by goat owners. They are becoming more common in some areas, however. (See page 200 for a typical barn record.)

With or without papers — registration certificates, pedigrees, show wins, barn or official test records, advanced registry certificates or stars — many people recommend that prospective goat buyers see the goat being milked, or better yet milk her themselves. For an inexperienced milker, the doe will probably be nervous and you might not get much milk. But it's better to get a lesson, even if the owner has to finish the job, than to get home and find out that you can't milk. Taste the milk to check it for off-flavors.

Assessing a Goat's Worth

You've found your goat, and you're ready to deal. Next question: how much is this goat worth?

Once again, there are no set answers. The price of goats generates as much heated conversation among goat people as anything else. Some plug for higher prices, some for lower prices, and there are good arguments on both sides. Goats have been sold for thousands of dollars, and of course, many more have been given away.

For the person whose primary interest in goats is an economical milk supply, there's a way to determine what an animal is worth. It involves guesstimating how much your home-produced milk will cost. But it will take some work, and it isn't foolproof.

Even a formula approach isn't really much help. It would be impossible to fill in the blanks of a formula in a book, because hay and grain prices vary widely from one area of the country to another, and so do milk prices. They also vary from year to year; in some places they can double after a drought or crop failure. Still, you need to start somewhere.

Basic Formula for Figuring Costs

- ◆ Add cost of grain times amount eaten per day and cost of hay times amount eaten per day.
- ◆ Multiply sum times 365 to get cost of feed per year.
- ◆ Divide number of pounds of milk produced per year by cost of feed per year to get annual feed costs per pound of milk.
- ◆ Multiply number of gallons of milk per year (pounds divided by 8) times local retail cost of goat milk to get annual retail value of your milk.
- ◆ Subtract feed costs from annual retail value of milk to approximate recoupable price to pay for goat.

An Example

Using the above formula in an example, let's say the local feed mill is selling Purina Goat Chow for $14.95 per 100 pounds, or 14.95 cents a pound. Although we found hay for our beef cattle for $1 per bale, we paid $2 per bale for some better hay for the goats; both lots were roughly 50-pound bales. If I feed my goat approximately 3 pounds of grain and roughly 4 pounds of hay per day (with some pasture), over the course of a year I'll use about 1,100 pounds of grain and 1,500 pounds of hay. (Note that hay consumption varies widely, depending on the type and quality, on how much pasture, if any, is used — and on how much the goats waste. Some people will use as much as 7 to 8 pounds of hay a day, per goat.) The grain will cost me, at current local prices, $165, and the hay will cost in the neighborhood of $60, for a total feed cost of $225 a year. If this goat meets my expectations and produces 1,500 pounds of milk, my feed costs for the milk will be just under 15¢ a pound ($225 divided by 1,500 pounds), or about $1.20 a gallon (15¢ times 8, the number of pounds in a gallon).

Goat milk in a health food store would cost me (if it were available) more than $6 a gallon, so the 180 gallons or so from my goat would have a retail value of almost $1,100! Ignoring incidental expenses, and certainly labor, I could pay $875 for the goat and get my money back within a year. The next year that could be considered clear profit, and if she has two kids, I'll do even better.

But wait a minute. Much as I prefer goat milk, what if $6 a gallon is too much to pay, so I opt to match the price of cow milk, which currently sells for $2.49 a gallon? Although we're comparing apples and oranges now, my goat milk would only have a value of about $450. I could figure on making $225 above feed costs.

On the other hand, I just heard from one of my nieces who is buying goat milk for her son who had problems with cow milk — and she's paying $3.50 a *quart!* Fortunately, he is thriving; unfortunately, I'm in Wisconsin and she's in Utah, so I can't help her out. But at $14 a gallon, she really should get a goat. Plugging that price into the above equation ($14 times 180 gallons per year, minus $225 in feed costs), her "profit" would be a whopping $2,295 a year.

Personal Considerations

There are many other factors to consider. I might not really use that much milk, or my usage requirements might not fit the goat's lactation curve. And of course, never forget that goats are living creatures, not machines. No one can predict if or when they'll get sick, or even die, or how much milk they'll produce.

If economical milk is your real concern, or if you enjoying playing with the spreadsheet program on your home computer, you can spend a lot of time juggling the numbers. By taking into account the life expectancy of the goat and the value of her kids during that period, as well as breeding, veterinary, and other expenses, you can get a fair idea of what a goat will be worth to you. You can also determine how much more you can afford to pay for a goat that produces 500 or 1,000 pounds more or less than the "standard" 1,500 pounds a year. You'll want to revisit this formula and these considerations when you start to upgrade, or when you wonder what effect reducing feed costs would have on your milk bill.

And if you're ever tempted to "go commercial," or wonder why goat milk can cost $14 a gallon in stores, plug in your labor! We won't even go into the cost of commercial equipment. And, of course, if goats are your business you'll have to account for everything from utilities and insurance to legal and accounting services. And don't forget taxes. Judy Kapture, an authority on commercial goat dairying, found that if all these issues are factored in, it typically costs around $20 per hundredweight, or cwt, to produce goat milk, and often much more. That's $1.60 a gallon or more.

Be Realistic

The real purpose of bringing all this up at this particular point is to demonstrate that anyone who thinks a decent dairy goat isn't worth more than $25 or $50 isn't being realistic. It costs more — in some places much more — than $100 just to raise a kid to milking age. If you find a seller who doesn't know that, or doesn't care, and will sell you a nice goat for little money, fine. But if a responsible breeder asks for a price in line with what the goat is really worth, don't go into shock. By the same token, if a seller wants to charge what you think is a ridiculous sum, go back over this chapter and assess your situation carefully one more time.

In the final analysis, chances are you'll get a goat just because you want a goat, and you believe that *any* goat is better than none. And as with a new car, you'll end up bringing home the one that fits not only your budget, but your personal tastes. If your new goat doesn't meet your needs, you might not keep her long, but I can make two guarantees: she will provide a learning experience, and you'll never forget her!

4

HOUSING

Many people, especially those who haven't had much experience with livestock, are prone to bring home an animal and *then* decide where and how they're going to keep it. This is definitely putting the cart before the horse. Fortunately, most people contemplating raising goats already have facilities that, with a little work, will serve as a shelter. If you're new to goats, you'd be well advised to learn something about them — not just from books, but from practical experience — *before* building any but the simplest brand-new facilities. A few years' experience will go far toward eliminating costly mistakes.

Most goats today are raised in what is called "loose housing." That is, instead of being confined to individual stalls, or kept stanchioned like miniature cows, the goats are free to move about in a common pen. While many cow dairies have been converting to this system to save labor, it makes even more sense for goats, because goats are social animals and need companionship; they are active animals and need exercise. Loose housing is a whale of a lot less work than individual box stalls or tie stalls. Loose housing obviously entails lower original cost in both construction time and materials. And it's more flexible, too. If you only have four stanchions, there's no way you can house five goats.

Ideal Housing

Goats are, in some ways, not awfully particular. They are commonly kept in garages and sheds, old chicken coops and barns, which may be constructed of wood, concrete, cement block, or stone. The floors may be wood, concrete, dirt, sand, or gravel. Although goat owners like to argue about the relative merits of each, goats do just fine in any of these types of housing.

Ideally, the goat house should be light and airy with a southern exposure. It should be convenient to work in, which means the aisles and doorways should be wide enough to get a wheelbarrow through without barking your knuckles; feed and bedding storage should be conveniently nearby; and running water and electricity should be available to make your work easier, more pleasant, and safer. Seems like most of the features of the "ideal" goat house are more for the benefit of the goat farmer than for the welfare and comfort of the goats!

Most goats are kept in a communal loafing area, or loose housing, rather than in individual pens.

An Ideal Starter Shelter

This simple, basic, attractive, economical, and practical goat shelter would be ideal for a beginner or anyone with two to three goats for the home milk supply. It uses standard dimension lumber, has an earthen floor, affords both protection from the elements and good ventilation, and can be built in less than 2 days for under $200 with all new materials.

Its shortcomings include lack of space for storing hay and grain, lack of water and electricity, and the milking stand being located in the shed. These could be overcome simply by expanding the structure or adding to it, and running water and electricity to it. But for our purposes, this structure works.

Hay is stored on pallets and covered with a tarp. Be sure the tarp is large enough to allow the first bales to be stacked on top of one edge, then tucked in along the two adjacent sides, with the fourth side firmly weighted down with cement blocks or something similar to prevent wind damage. Besides its minimal cost, this type of storage "shrinks" as the hay is used, so you don't pay for a structure that is half empty half of the time.

Grain is kept in a metal garbage can in an adjacent garage.

In good weather the goats are milked outdoors. Milking equipment is washed and stored in the kitchen.

If you decide to keep more goats, this structure could be expanded or used as-is for feed storage, or kid raising, or it could be converted to a milking parlor. But again, this type of shelter is all you need to get started. And with the experience it will give you, you'll be better able to design your own "ideal" goat facilities. How about a brick or fieldstone or log goat house? Let your imagination, budget, and experience be your guides.

As for the goats, they do not have to be kept warm even in northern climates if they've been conditioned to the cold through the fall. But in any climate, their housing must be dry and free from drafts. Goats are very susceptible to pneumonia.

Buildings should be whitewashed or painted white. This will make the building more attractive and pleasant to work in, for you and the goats, and light colors tend to discourage flies and other pests.

Lighting the Goat House

Lighting can be used both to increase fall production (20 hours/day starting in September in Wisconsin's latitude, for example) and to induce spring breeding (14 hours/day starting March 1, to simulate the shorter days of fall).

This chapter includes some suggested floor plans. You'll want to adapt them to your own particular circumstances. After a few months of doing chores you'll probably want to make some changes based on your building, your animals, and the way you do things. Again, don't get too fancy or spend too much money right away if you are a novice and don't have any experience with goats.

Flooring

Is there an ideal flooring material for the goat house? Your situation will determine your preference, but the floor and its proper maintenance will contribute a great deal to the health and comfort of your goats.

Wooden Floors

Wooden floors such as those found in brooder houses or other poultry buildings can be warm and dry if the rest of the structure is snug and tight. But wood rots. This means highly absorbent bedding should be used, and it should be changed frequently.

The most absorbent bedding is peat moss, which can absorb 1,000 pounds of water for each 100 pounds of dry weight (hundredweight, or cwt), far more than any other bedding material. But if you have to buy it, it's expensive, so few goat owners use it.

Chopped oat straw rates second among commonly used materials, but it only absorbs 375 pounds of water per 100 pounds of dry weight. Note that this is *chopped* straw. Long straw will only absorb 280 pounds per hundredweight (cwt). Wheat straw is somewhat less absorbent than oat straw.

So wooden floors are obviously not the most desirable for animals like goats, where large quantities of wet bedding will accumulate. If you build a new structure for goats, don't put in wooden floors. But if you already have a building with wooden floors, there's no reason not to use it.

Construction Tip

Use screws rather than nails when you build. Screws are stronger, and they come out much more easily when remodeling time comes.

Concrete Floors

According to many experienced goat raisers, concrete floors are only somewhat less desirable than wooden floors. Concrete is cold. The urine cannot run off, so concrete floors require a great deal of bedding. I don't consider this a serious drawback unless bedding is very expensive and/or you don't have a garden to use it on. Many people prize the used bedding for use in their fields and gardens or compost bins.

Goat manure and cow manure are quite different. Cow manure is extremely loose and liquid, certainly in comparison with nanny berries. Those neat little compact balls bounce, and in this context bouncing is preferable to splashing. With a little fresh bedding to keep the top surface clean, goat litter can accumulate to a considerable depth and still be much less offensive than a cow stall.

The problem with concrete floors is that while the surface might be quite clean and dry, the bottom layers can be a swampy morass. Deep litter does not signify a sloppy goat farmer; quite the opposite. For goats, the deeper the better, certainly on concrete floors, and especially in winter. The lower layers will actually compost in the barn if they aren't too wet, not only helping to warm the goats' beds in the same way the old-fashioned hotbeds warmed early garden seeds, but also hastening its use in the garden. Some people even use compost activator on the bedding to speed up the bacterial action. Such deep litter is warm, and quite odorless — until you clean the barn, that is.

Goats prefer less bedding in the warmer months, but even then concrete floors require special management. Concrete is always relatively cool, and hard, and there have been reports of rheumatism-like problems arising among goats on hard surfaces. Sleeping benches — simple raised wooden platforms — can be used both winter and summer, and the goats are enthusiastic about these regardless of floor type.

Concrete floors have the great advantage of being easy to get really clean, which can be particularly important in the summer when deep litter might not be so desirable. In fact, Harvey Considine, one of the best-known and most highly respected goat breeders in the country, prefers concrete floors, primarily because of sanitation. (Of course, he has several hundred goats and uses a skid steer loader to clean his pens.)

Some years ago there was a lady who kept goats in what was practically the center of town, as the suburbs grew up around her. She kept the animals on concrete floors, without bedding. The urine drained away, and the droppings were swept up daily. The place was spotless, there was never a complaint from the neighbors, and the only waste disposal was a daily coffee can of nanny berries that went on the rose bushes.

Dirt and Gravel Floors

I have had concrete floors, and I liked them. But other people have actually torn out concrete to install what they consider the ideal: dirt or gravel floors. That's what I have now. I like that, too, and I have plenty of company.

Earthen floors are the easiest to maintain. Excess urine soaks away, and less bedding is needed. This is an important consideration if bedding is expensive and you don't have gardens to use it on. Still, there will be plenty of bedding for the compost pile. Soil is also warmer and more comfortable for the animals than concrete, especially if only a small amount of bedding is used.

Insulation

We'll assume that the roof doesn't leak and that drafts aren't getting into the house through cracks in the walls and around windows and doors. We do want ventilation, but not drafts.

Insulation often isn't necessary, but it can be highly desirable. It can eliminate condensation, help keep water from freezing, and make the goat barn a more comfortable place for you to work. But you must take special precautions to protect insulation from the goats. Goats will chew, eat, or smash any number of wall materials. Plywood, plasterboard, and the like won't last more than a couple of days. Use stout planks or cement wallboard instead.

Whether or not you insulate, *never* use a plastic moisture barrier in a goat shed or barn. The plastic won't keep the goats any warmer, and it will cause condensation and humidity problems.

Size Requirements

The size of the building depends on several factors, including herd size, climate, size and availability of pasture and space for exercise, feed and bedding storage requirements, and whether you need, want, or can manage a separate milking parlor.

For example, standard recommendations range from 12 to 20 or 25 square feet per animal. In warm climates, you can go with the lower figure, because animals spend more time outside. If there is a sizable pasture or exercise yard and the barn is used mainly as a dormitory, you could also get by with a smaller figure.

But adequate space for the animals themselves is only one consideration. If you're working with minimum space requirements, don't forget that the addition of kids will require more room.

And then too, there is always the possibility that your herd will increase. Goat herds have a way of doing that, even on well-managed and very small places!

An Idea for Doors

An overhead garage door works well in a loafing shed. It won't become obstructed by bedding and manure like swinging or sliding doors will. It can be opened to whatever extent weather conditions dictate. Used garage doors are often available for little or nothing.

You'll need space to store hay, bedding, and grain, and more likely than not the milking will be done in the barn too. (Technically, it should be in a separate section, away from dust, manure, and odors, but most people with only a few goats find this hard to justify. For them, milking outdoors, in nice weather, can be an acceptable substitute, but milking in a barn aisle is common.)

In other words, if you grow your own hay and straw or buy it off the baler when it costs less and plan to store a year's supply, you'll naturally need more room than if you intend to bring home a week's supply at a time. If you spend more money to have a larger building, you can hope to save some by buying feed and bedding in large quantities. Note, though, that if you stock up on hay and straw during the harvest, roughly half of this will be gone by kidding season. Consider planning a combination hay-storage and kid-raising area.

Here again, if you already have facilities that aren't perfect, don't panic. If there simply isn't room for hay in the building you'd like to use for goats, it can be stored elsewhere. It will involve more labor, but it won't keep you from raising goats. Grain can be stored in garbage cans in outbuildings, if necessary. Milking elsewhere can be extremely inconvenient however, especially in inclement weather. Unless an alternate shelter is very close to the goat shed, try to incorporate milking space into the main building.

Grade A dairies, cow or goat, must have a separate room for handling milk. A separate milking parlor is advisable to eliminate dust and barn odors. But for the backyarder or homesteader with a few animals that are kept clean, milking in an aisle is acceptable, and far more convenient. Also, a milking bench isn't a necessity, but it will contribute greatly to the ease of milking, and will result in a superior product.

Note on Grade A Dairies

Regulations governing grade A dairies (goat and cow) vary by state and, in some cases, by locality. Contact your local county extension agent for information about the requirements in your area.

The Manger

Never just throw hay or grain on the ground — use a manger to feed your goats. Grain is often fed in mangers, since most goats won't get their allotted ration during milking. Greed can cause problems, unless some means of fastening the animals into the manger is devised to prevent bossy does from taking more than their share. Hay can be fed free choice in the same manger; the goats can come and go as they please and eat as much as they want.

There are many styles of mangers, but the best ones consider the nature of the goat. Goats are notorious wasters of hay, and this is the main factor to consider when designing and building a manger. It should also be difficult for goats to climb into or otherwise contaminate, and of course it should be easy for the goats to eat out of and for you to clean and fill.

The keyhole manger, a longtime favorite, has the shape of the classic keyhole, with a 7-inch diameter hole at the top and a 4-inch wide slot coming down from that. The doe must reach up to get her head through the hole, and then slide her neck down the slot. If you want to constrain a goat in this device, perhaps to trim hooves or even for milking, a simple latch to keep her neck in the slot will do it.

Based on his half a century of experience with goats — as many as 1,000 at a time — breeder Harvey Considine has devised a new version of the keyhole manger. It has two unusual features that deal with goats' habits.

First, the goats eat standing with their front feet on a "step" a few inches above the floor level. And to get at the feed they must put their heads through openings made of slats or bars set at an angle of 63 degrees. The openings range from about 4 inches for small kids, to 5½ inches for mature does, to 7 inches for large bucks. The animals must then reach *down* to get the feed.

The angle is important. The goats can see the hay, and get at it by tipping their heads. Ordinarily, goats would grab a mouthful of hay and back off, dropping half of it on the floor — and then refusing to touch it again. But the angled bars, Harvey says, do away with this problem: "If they pull their head out quickly, they will almost certainly give themselves a nasty clip on the ear because they forgot to turn their head. Within a few days you will find that the old grab-a-bite-and-slug-your-neighbor syndrome is eliminated. They will soon just line up and stay with their heads in the manger until they are done eating. It is a beautiful sight." Indeed! Especially if you're paying several dollars a bale for hay!

Goats are notorious for wasting hay. A well-designed manger can minimize this problem.

Another Feeder Idea

Sections of PVC pipe, cut in half lengthwise, make good feeders for grain or loose minerals, or even milk. A 3- to 4-inch diameter pipe works well for kids; mature animals need about 15 inches in diameter. Make the feeders as long as you need. Block the ends with PVC caps, and hang the feeders with metal strapping. The PVC is easy to clean.

You can use 1×4 boards for the slats, but these obstruct the goats' view of the feed and aren't very durable. Instead, Harvey recommends steel strapping, ⅛ inch thick, 1½ to 2 inches wide, and about 24 inches long. The bottom feed trough can be any width, but a goat won't be able to reach more than about 16 inches. These should be about 42 inches high on the goats' side to keep them from jumping into the manger.

The manger is best built as a fence or divider inside the barn so you can fill and clean it from the aisle. It can be built along a wall, of course, but that will make your chores more difficult.

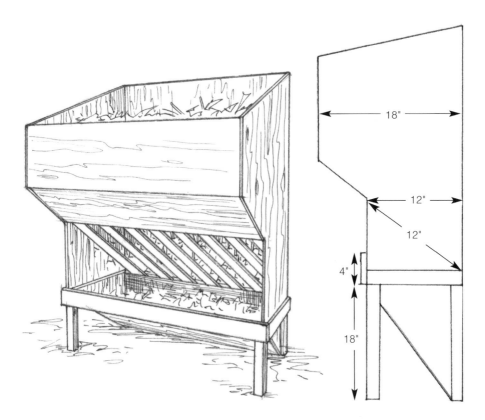

This type of hay manger is simple, yet functional. Make it as long as necessary for the number of goats you're feeding, allowing 18 inches or more per animal.

Gates and Latches

Gates and latches are important in goat houses. Gates should be sturdy, for goats love to stand on things with their front feet, and gates are the favored place to do this standing. With deep litter, the gate should swing out of the pen, which is a good idea in most cases anyway. Make sure the gate is wide enough to get through with a wheelbarrow or whatever you'll be using at cleaning time. Sliding gates are very nice, but more difficult to build. As to the ideal height for fences: goats vary widely in jumping ability, or perhaps desire. Contented goats are less likely to leap fences of any height. But if a deep litter system is used, remember that a 4-foot fence in October might only be 3 feet high a short time later! (And naturally, as the floor goes up, the ceiling comes down, an important consideration if it's already low or if you're tall.) Since most goats are Houdinis when it comes to unlocking latches, pay special attention to those. A double-jointed eye hook is a good choice.

Sturdy gates and secure latches are of special importance in goat barns. Goats are amazingly adept at opening ordinary hooks and latches.

Other Considerations

While some people, goat raisers or not, demand more luxuries and conveniences than others, of particular interest are running water and electricity. Storage space is also a concern.

Utilities

Water piped to the barn can save countless minutes, which on an annual basis amounts to hours or even days. The goats are more likely to have a continuous supply of fresh water if you don't have to lug it a long distance, and from that standpoint alone the plumbing can be worthwhile. A hose might work in summer or in a warm climate, although it's an unsightly nuisance and the water can get quite hot. Where freezing occurs, buried pipe and a frost-proof hydrant can almost be considered necessities.

Electricity is obviously a boon when you have to do chores before or after the sun shines. Trying to milk or deliver kids by flashlight is hectic, and lanterns can be dangerous as well as a bother. Moreover, eventually you'll want electricity for clippers, disbudding irons, heat lamps, and perhaps for a stock tank heater to keep drinking water from freezing, as well as other possibilities. Some people say a radio tuned to a classical music station makes their goats give more milk.

Caution

Be certain that wiring and cords are not exposed for goats to chew.

Storage Space

You'll want storage space, of course. How much you need depends on the type of operation you have. There should at least be room for a pitchfork that can be kept out of harm's way, hair clippers, hoof trimming tools, brushes, disbudding iron or caustic, and a medicine cabinet. Provide a place for a hanging scale, and make sure to keep the milk records where the goats can't nibble on them! (This is the voice of experience speaking.)

Milking equipment must be stored somewhere cleaner than the barn, of course. Most home dairies use the kitchen as a milkhouse. The ideal milkhouse — well ventilated, with hot and cold running water, rinse sinks, floor drain, and impervious walls and ceiling — is nice, but a bit much to expect for a dairy with only a few goats. The kitchen works just fine for most people. That's where the utensils and strainer pads are kept, the milk strained and cooled, and all milking utensils washed.

A Tip for Keeping Facilities Clean

Scrub down mangers, walls, and concrete floors with a bucket of hot water containing 1 cup each of bleach and cider vinegar. Dirt floors should be limed.

Final Thoughts

Facilities for dairy goats need not be elaborate or expensive. But because goats are dairy animals, you'll want to keep them and their surroundings as clean as possible. Plan quarters that are easy to keep clean, that are pleasant for both you and the goats to be in, and that will contribute to the health and well-being of your herd. In the end, the design of your barn depends on your building site, budget, individual situation, and personal preferences There is no "best" plan for everyone. If a design meets your needs, it's a good one.

Sample Floor Plans

18 x 18 Goat Barn

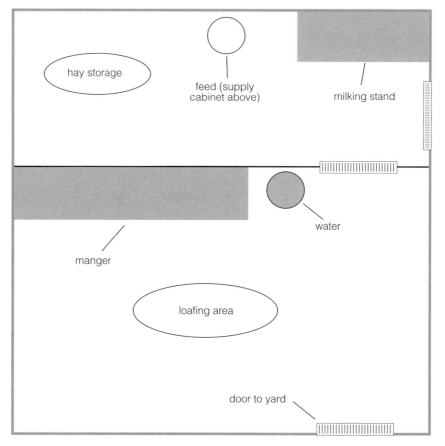

This is a good basic floor plan, showing the fundamentals of a goat barn. Grain is stored in a metal garbage can with a tight lid. Many alternative arrangements are possible, including locating the water outside the loafing area to help reduce contamination, and providing a wider door to the yard to prevent bossy does from blocking the entrance.

8 x 15 Shed

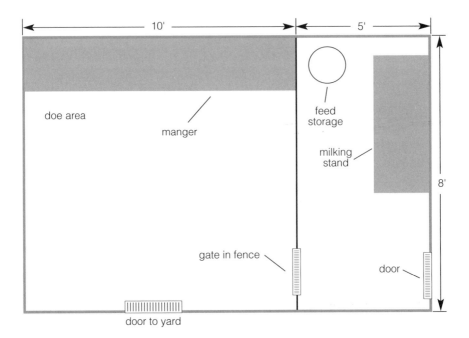

Although a very small shed is workable, it can hamper efficiency. This plan would require carrying hay from an outside storage area all the way through the shed to reach the manger. Sliding doors and a folding milking stand (see page 186) are especially valuable in cramped quarters like this.

Helpful Hint

Hang a toilet scrub brush near the water faucet, and clean water buckets every time you fill them.

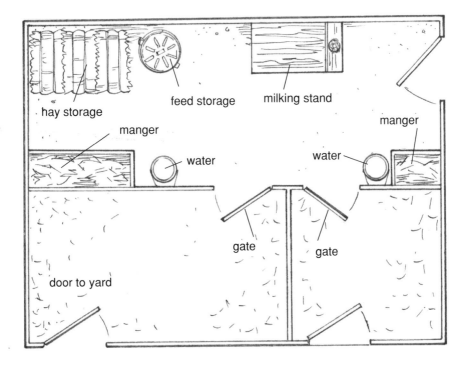

hay storage

feed storage

milking stand

manger

manger

water

water

gate

gate

door to yard

Barns should be designed for efficiency. They should be easy to clean and keep clean. The layout should eliminate unnecessary motion. Function is more important than appearance.

In this design the mangers and water buckets are outside the pens, and the doors and gates swing into the pens. These are personal choices. Some people wouldn't want the two pens shown here. But if you only have a few goats and one shed and want to keep a couple of kids, this arrangement could work just fine. (The smaller area might also be used as a kidding pen and/or for hay storage.)

FENCING

Fencing is probably more important — and more difficult — with goats than with any other domestic animal. Goats will jump over, crawl under, squeeze through, stand on and lean against, and circumvent any boundary that is not strictly goat-proof in any other way they can discover or invent. In addition, fencing often serves a second purpose — to protect the goats from stray dogs and other predators.

How Much Is Necessary?

With goats, a little fencing goes a long way. In most cases, if you only have a couple of goats you won't want to think in terms of "pasturing" to any great extent. Goats won't make good use of the usual pasture plants, grasses, and clovers. They prefer browse: trees and shrubs and brush. Goats that are fed at the barn will probably ignore even the finest pasture, although they'd be delighted to get at your prize roses, specimen evergreens, and fruit trees. For many people, protecting valuable plants like these is the main reason for good fences! Goats also like to jump on cars, so make sure vehicles and goats are kept apart.

We'll talk more about pastures and pasture fencing later. For now, let's focus on the exercise yard. A small, dry, sunny yard adjacent to the barn is all you need, ordinarily, and you'll probably want one of these even if you pasture your animals. The exercise yard fence will take more punishment than the average pasture fence, because the goat confined to the smaller space will have more time and opportunity to investigate and beat on it. The cost per running foot will be higher in the yard, but the amount of fencing used is much less.

What Kind of Fence?

Good fencing is obviously a necessity for goats. And while fences do require an investment, the newer types make it much easier and more economical to allow goats access to larger yards or pastures. First, are the fences to avoid, followed by the ideal types and some practical alternatives.

Types of Fencing to Avoid

For various reasons, these fences are not recommended.

- ◆ **Woven wire.** This type, also known as field fencing, is less expensive, but has drawbacks. If your goats have horns, they'll put their heads through the fence, then be unable to get free. Worse, they'll stand on the wire, or lean against it, until it sags to the ground and they can nonchalantly walk over it. Even with close spacing of posts and proper stretching — a crucial part of building this type of fence — woven wire will soon sag from the weight of goats standing against it and will look unsightly, and eventually be useless. However, woven-wire fencing can be useful when combined with electrical fencing (see page 71).

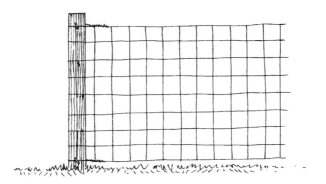

Woven-wire field fencing is relatively inexpensive, but goats can easily ruin it by standing on and leaning against it. It can be used with wood or metal posts.

+ **Rail fence.** For many people the picturesque board or rail fence comes to mind first, but it won't work for goats unless it's all but solid. They can slip through openings you wouldn't believe. Don't take the chance.
+ **Barbed wire.** It's awfully ugly stuff around tender-skinned, well-uddered goats. And it doesn't impress them anyway. Some people use it, often because they feel it deters predators. Far more would suggest getting rid of it.
+ **Picket-style.** Another nasty trap for goats. With a picket-style fence, there's a very real danger that a goat will stand against the fence with her front feet, slip, and impale her neck on or between pickets.

Useful Types of Fencing

Here are better, safer options that work.

+ **Chain-link.** This is the ideal goat fence for a small place. However, like most ideals, it may simply not fit the budget.
+ **Stock fencing.** A very good and somewhat less expensive alternative to chain-link, stock fencing, often called hog or stock panels, consists of welded steel rods. The panels are usually 16 feet long and come in several heights. For goats they should be at least 4 feet high. Attach stock fencing to regular steel or wooden fence posts; or connect the panels with small cable clamps.
+ **Sheep stock fencing.** Regular stock fencing has one problem: horned goats can get their heads trapped between the rods of the panels. A newer and better version has smaller spaces (about 4 inches) between the rods. This makes it more expensive but much safer for your goats. This type of fencing was designed for sheep, and, unfortunately, it isn't available everywhere.
+ **Electric fencing.** This should be used much more than it is for goats. The goats have to be trained to respect it, but once they know what happens when they touch it, it's possible to fence even large areas at low cost. Train the goats in a small area. Until they get zapped once or twice they'll be crawling under, jumping over, and just plain busting right through.

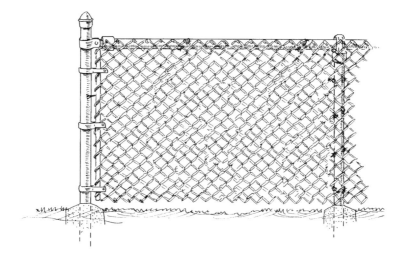

Chain-link fencing is ideal for goats, but it's expensive.

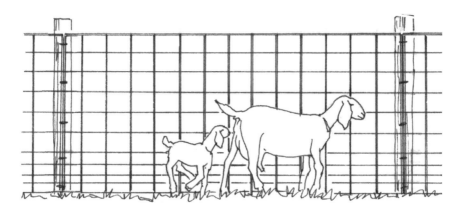

Stock panels are made of ¼-inch welded steel rods, which makes them sturdy and ideal for goat pens. The standard length is 16 feet; the 52-inch high ones work well for goats. A recent improvement features panels with 4-inch spaces, which eliminates the problem of horned goats getting their heads stuck in the fence.

You might want to use sturdier (and more expensive) fences for smaller yards, but electric fences are ideal for larger areas such as pastures.

You can also use electric fencing in tandem with woven-wire, or field, fencing. Place a strand of electrified wire just inside the woven-wire, at about nose height. This combination makes a very good goat barrier: the electric fence keeps the goats from reaching the field fencing, and the field fencing offers more security than the hot wire alone.

An electric fence carries a pulsating (not steady) current provided by a fence charger, which can operate on household current or a battery. Solar-powered models are available. The fence must be properly grounded and kept free of weeds, which will short out the current, making the fence ineffective.

A seven-wire, high-tensile electrified fence like this should do an excellent job of keeping goats where they belong.

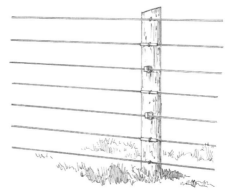

◆ **New Zealand–type.** A recent addition to the possibilities, New Zealand–type fencing is powered by a 12-volt battery and an energizer that, when properly grounded, intermittently sends out 5,000 volts. It uses fiberglass posts, except on corners. New Zealand fences, originally designed for intensive grazing systems, have enjoyed a rapid rise in popularity and led to many innovations, including portable electric fence.

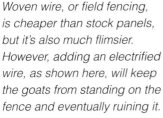

Woven wire, or field fencing, is cheaper than stock panels, but it's also much flimsier. However, adding an electrified wire, as shown here, will keep the goats from standing on the fence and eventually ruining it.

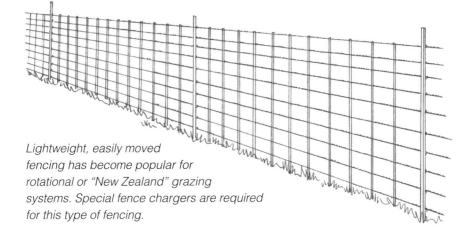

Lightweight, easily moved fencing has become popular for rotational or "New Zealand" grazing systems. Special fence chargers are required for this type of fencing.

6

FEEDING

No aspect of goat raising is more important than feeding. You can start out with the very finest stock, housed in the most modern and sanitary building, but without proper feeding your animals will be worthless. And feed is the goat owner's biggest expense.

The proper feeding of goats requires special emphasis for several reasons. High on this list is the fact that many people who start to raise goats have little or no experience with farm animals. Feeding goats is a lot different than feeding cats or dogs or parakeets. Goats are ruminants, which affects their dietary needs. And unlike the average cat or dog, goats are productive animals, which puts additional strain on their bodies and requires additional nutriment.

The bulk of a goat's diet consists of forages: green plants or hay. But this doesn't mean that just any old grass will provide the nutrition a healthy, milk-producing or pregnant goat requires. So we'll take a look at some of the more common pasture plants and what every goat owner should know about hay.

Browse and hay do not provide all the vitamins and minerals the productive animal needs, however. The *concentrate*, or grain ration, provides these. We must know at least the basics of this, too.

But before we talk about specific feeds, let's weigh some of the considerations involved in selecting a feeding program.

Proper feeding of goats — for their well-being and for milk production — requires more than turning them out into just any old grassy area. This chapter explains why, and how to do it right.

The Long and the Short of It

A discussion of feeds can be very long, or very short. It can be short if you buy good hay and a commercially prepared grain ration and follow the directions on the label. It will have to be long if you grow hay and mix your own concentrates because that will require at least an awareness of the elements of nutrition, physiology, bacteriology, math, and more.

There is no middle ground. You can't, for example, arm yourself with a "goat food recipe" and do a good job of feeding.

Most goat owners, especially beginners, buy hay from farming neighbors and ready-mixed concentrates from the feed store, so this should be a short chapter. But it's not — not so much because you might want to grow your own feed, but rather because an understanding of what's involved in those "off-the-shelf" feeds will make you a better goat breeder and give you a greater appreciation of your goats' needs.

We'll discuss pasture, hay, and grain. But to understand their importance, we should know something about the animal's digestive system.

The Digestive System

People familiar only with human diets and perhaps those of dogs and cats should especially examine the process of rumination, because goats are ruminants. Like cows and sheep, goats have four "stomachs."

The process of rumination serves a very definite purpose and has an important bearing on the dietary needs of the animal. Ruminants feed only on plant matter that consists largely of cellulose and other carbohydrates and water, making adaptations in the structure and functioning of the stomach and intestines necessary. We commonly speak of "four stomachs," but in reality the large rumen (or paunch), the reticulum, and the *omasum* ("many plies") are all believed to be derived from the esophagus, while the fourth stomach, the *abomasum*, is the true stomach and corresponds to the single stomach of other mammals.

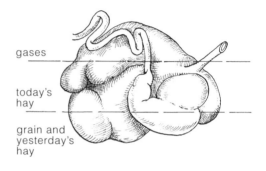

gases

today's hay

grain and yesterday's hay

The top of the rumen is filled with gas, and the middle contains recently eaten hay, which floats on the bottom slurry of yesterday's hay, grain, and fluid.

Vast numbers of protozoans and bacteria live in the rumen and reticulum. When food enters these "stomachs," the microbes begin to digest and ferment it, breaking down not only protein, starch, and fats, but cellulose as well. The larger, coarser material is periodically regurgitated as the cud, rechewed, and swallowed again. Eventually the products of the microbial action (and some of the microbes themselves) move into the "true" stomach where final digestion and absorption take place.

No mammal, including the goat, has cellulose-digesting enzymes of its own. Goats rely on the tiny animals in their digestive tracts to break down the cellulose in their herbivorous diet. You might say you're feeding the microbes and the microbes feed the goat, for without them the grass and hay would have no food value.

These tiny organisms, conditioned as they are to a given diet, can't cope well if their diet alters. The result of a diet change is usually a sick goat. Therefore, make any feed changes gradually. Many a goat raiser has fed a goat an armload of cornstalks salvaged from the garden after harvesting sweet corn, and when the goat gets sick or dies, the cornstalks get the blame. In reality, the problem was overload. Feed such delicious things sparingly, along with the regular diet, and everybody — protozoans, bacteria, goat, and you — will be much happier.

Let's back up a bit to take another look at these stomachs, for not only are they of obvious importance to the goat: the goat owner (or at least the kid raiser) has some control over their development.

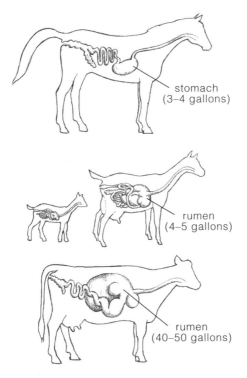

stomach
(3–4 gallons)

rumen
(4–5 gallons)

rumen
(40–50 gallons)

As an illustration of why the feeding requirements of ruminants differ from single-stomach animals, note that a horse's single stomach holds 3 to 4 gallons, while the rumen of a well-developed but much smaller goat has a capacity of 4 to 5 gallons. A cow's rumen can hold as much as 40 to 50 gallons!

Developing the Rumen

The rumen and reticulum together occupy about 30 percent of the stomach space of a young, milk-fed kid. At maturity, a well-developed doe has a rumen that occupies 80 percent of the stomach space and a reticulum that takes up 5 percent. The goat must have a well-developed rumen to function properly and requires a bulky diet to keep the rumen working properly. The rumen does not increase in size without proper stretching or development; therefore, early feeding of roughage is essential. And for these reasons, hay (or other roughage) forms the basis of the goat's diet.

How a Kid Drinks

Watch a newborn or very young kid sucking. She stretches her neck out to get her milk. Due to the stretching process the milk goes past a slit in the esophagus, by-passing the first two stomachs and ending up in the omasum. Here it is mixed with digestive fluids and is passed on to the fourth stomach, or abomasum.

Contrast this with a pan-fed kid, especially one fed only two or three times a day instead of four or five, and who is therefore more hungry and greedy. It must, first of all, bend down to drink rather than stretch upward. Some of the milk slops through the slit in the food tube and falls into the first stomach, the rumen, where it doesn't belong. There is nothing else in this compartment, since milk is the only feed consumed. There is no bulk. Gas forms, and scours are likely to result. (On the other hand, pan feeding is much easier than bottle feeding. And pans are easier to keep clean and sanitary than bottles and nipples, a fact that may offset any potential downside.)

The good goat raiser strives to keep milk out of the rumen by proper feeding. Moreover, the breeder will work to develop the rumen and reticulum the way they should be developed by encouraging the kid to eat roughage at an early age.

If you watch a mature doe eat, you'll see that she takes little time to chew. She draws in her neck to swallow, allowing the food to slip through the slit in the esophagus to the rumen. A slight fermentation begins as the microbes go to work. When at leisure, the doe regurgitates some of this material and "chews her cud." To swallow the now thoroughly masticated food, she extends her neck, and the cud goes to the third stomach, or omasum. When she drinks from a pail of water, she extends her neck to the far side, ensuring that the fluid goes to the omasum where it belongs, not to the rumen.

Feeding for Milk Production

In addition to the special needs of the goat relative to rumination, it's important to feed her as a *dairy* animal. Production of milk requires more protein than would be needed just for body maintenance, for example. So a milking doe is fed a ration of at least 16 percent protein, while a dry mature doe or buck will do well on 12 percent. Protein is expensive, and any excess is just wasted. You want to make sure the diet has enough, but not too much. Dairy animals have a greater need for calcium and certain trace minerals as well.

Bulk and the British Postwar Experience

Excessive feeding of milk and concentrates to young goats apparently prevents full development of the rumen; advocates of early weaning agree.

In his book *Goat Husbandry* (London: Faber and Faber, 1957), David Mackenzie maintains that bulk is necessary for good milk production. He points out that milk production in British goats dropped by 12 percent just in the 4 years following postwar "derationing" of animal feedstuffs in 1949.

When concentrates were rationed during the war years, the official concentrate ration for a milking goat was adequate if she had plenty of bulk food such as hay and roots. The allowance for kids and young stock was much more restrictive, and milk for kids even more so. Despite this, Mackenzie's charts show a steady *increase* in milk production during rationing, based on about 3,000 records from the British Goat Society, and a dramatic *decrease* in milk production after rationing was lifted.

Basic Nutritional Requirements

It would be very helpful to think in terms of minimum daily requirements for humans, which most of us are familiar with nowadays. Goats, too, have minimum daily requirements. Remember this, and you'll be less tempted to stake the animal in a brush patch and assume she's "fed" just because she filled her belly. She has no more nourishment in that situation than you would if you lived on candy bars and soda pop.

Unfortunately, there are few definitive minimum daily requirements listed for goats. With several notable and recent exceptions, very little research has been done with goats, certainly in comparison with the more economically important livestock such as cattle and hogs. This is now changing, but meanwhile, we can look at nutrition in general and make some assumptions based on what is known about other ruminants.

Water

Like any other livestock, goats should have constant access to fresh water. But all feeds contain water, too. Water is vital to life of course, but it's also important in feed formulations because the quantity of water in various plants affects their place in the ration. Dry grain, for example, might contain 8 to 10 percent water. Green growing plants might contain 70 to 80 percent water. An animal fed succulent plants ingests an enormous amount of water in order to get nutrients.

Carbohydrates

Of the plants' dry matter, about 75 percent is carbohydrates, the chief source of heat and energy. These carbohydrates include sugars, starch, cellulose, and other compounds.

The sugars and starch are easily digested and have high feed value. Cellulose, lignin, and certain other carbohydrates are digested only with great difficulty and therefore it takes energy to digest them: their feed value is correspondingly lower. (This is one reason goat raisers prefer "fine-stemmed, leafy green hay." The fine stems means less lignin and hard-to-digest materials.)

If you buy feed, the feed tag on the sack will have the carbohydrates divided into two classes: crude fiber (or just plain fiber) and nitrogen-free extract. Nitrogen-free extract is the more soluble part of the carbohydrates and includes starch, sugars, and the more soluble portions of the pentosans and other complex carbohydrates. It also includes lactic acid (found in milk) and acetic acid (in silage). Oddly enough, nitrogen-free extract also includes lignin, which has a decidedly lower feeding value than cellulose.

Fats

Feed tags also list "fat," which actually includes fats and oils. They're the same except that fats are solid at ordinary temperatures while oils are liquid. In grains and seeds, fat is true fat. In hays and grasses, much fat consists of other substances. Many of these are vital for life, including cholesterol, ergosterol (which can form vitamin D), and carotene (which animals can convert into vitamin A). Note that these are all fats from plant, not animal, sources.

Proteins

The proteins and other nitrogenous compounds are of outstanding importance in stock feeding. Proteins are exceedingly complex, each molecule containing thousands of atoms. There are many kinds of proteins, some more valuable than others. (Livestock feeders speak of the "quality" of protein.) All are made up of amino acids, of which at least twenty-four have been identified, and protein must be broken down into amino acids before it can be absorbed and utilized by the body. Because they can combine like letters of the alphabet, there could be as many proteins as there are words in the dictionary.

The protein in plants is concentrated in rapidly growing parts (the leaves) and the reproductive parts (the fruits, or seeds). In animals, protein comprises most of the protoplasm in living cells and the cell walls, so it's important for muscles, internal organs, skin, wool or hair, feathers or horns, and it's an important part of the skeleton.

Protein, or crude protein, includes all of the nitrogenous compounds in feeds. It's of extreme importance to the animal caretaker, and it's obviously essential for life, but needs vary among classes of animals. Protein requirements are higher for young and growing animals, reproduction, and lactation. And because protein is the most expensive portion of livestock feed, you won't want to offer more than necessary.

Minerals

Ash indicates the mineral matter of the ingredients. Minerals in plants come from the soil, but the mineral content of animals is higher than that of plants. Calcium and phosphorus are particularly important since they are the chief minerals in bone and in the body. The body contains about twice as much calcium as phosphorus, and the proper balance is important.

Other minerals are needed in trace amounts, but they are vital. Iodine, for example, prevents goiter; iron is important for hemoglobin, which carries oxygen to the blood; copper, which is a violent poison, is also a vital necessity in trace amounts, as a lack of iron, copper, or cobalt can result in nutritional anemia. We'll come back to minerals later.

Total Digestible Nutrients

Net energy values of livestock feeds are expressed in *therms* instead of calories. Since a therm is the amount of heat required to raise the temperature of 1,000 kilograms of water 1°C, one therm is equal to 1,000,000 calories.

Nutrients are constantly being oxidized in tissues to provide heat and energy. This oxidation maintains body heat and powers all muscular movements. Since the digestion of roughages requires more energy, it follows that 1 pound of total digestible nutrients, or TDN, in roughages will be worth less than 1 pound of TDN in concentrates, which will not use up so much of its energy just being digested. Total digestible nutrient refers to all of the digestible organic nutrients: protein, fiber, nitrogen-free extract, and fat. (Note that fat's energy value for animals is approximately 2.25 times that of protein or carbohydrates.)

Digestible, of course, refers to nutrients that can be assimilated and used by the body. For this reason protein or crude protein is different from digestible protein. Digestible nutrients are determined in the laboratory by carefully measuring the amount of feed consumed and analyzing its content, and then analyzing the waste products. (Animal feces are largely undigested food, in contrast to human feces, which have a larger proportion of spent cells and other true "waste.")

Vitamins

Another important consideration in feeding is vitamins. Vitamins were largely unknown before 1911, and there is still more to learn about them. But as of now, the only two of any consequence to goats are vitamins A and D.

Vitamin A
Vitamin A is of prime importance to dairy goats because it's necessary for growth, reproduction, and milk production. It is of less importance in maintenance rations. Vitamin A is synthesized by goats that receive

carotene in their diets. The chief sources of carotene are yellow corn and leafy green hay. Common symptoms of vitamin A deficiency are poor growth, scours, head colds and nasal discharge, respiratory diseases including pneumonia, and blindness. A severe lack of vitamin A prevents reproduction or produces weak (or dead) young at birth.

Vitamin D

The other important vitamin for goats is vitamin D. As with other animals, lack of this vitamin causes rickets, weak skeleton, impaired joints, and poor teeth. Vitamin D is necessary to enable the body to make proper use of calcium and phosphorus. The best and chief source is sunshine, but it is also available in sun-cured hay.

Other Vitamins

The B-complex vitamins are manufactured in the rumen, so the feeder has no concern with them directly. Vitamin E seems to have no special application to goats. Vitamin C is synthesized. (Only humans, monkeys, and guinea pigs lack the ability to manufacture vitamin C.) Vitamin K is also synthesized.

Formulating a Goat Ration

With this very brief background of a goat's nutritional requirements, we can begin to formulate a ration. Since we now know how rumination works, and how important roughage is to that process, we'll begin with roughage. We'll then consider grains before developing some sample rations.

Roughage

Roughage can be green, growing plants, including grasses, clovers, and the trees and shrubs goats eat. It can be plants in dried form, called hay. There are two types of hay: legume hay, made from alfalfa or clover; and carbonaceous hays made from timothy, brome, or other grasses. Corn stover (dry cornstalks), silage (fermented corn plants or hay plants), comfrey, sunflower and Jerusalem artichoke stems and leaves, and root crops such as mangel beets, Jerusalem artichokes, carrots, and turnips are also considered roughages.

clover

Green Forages

Green forages are rich in most vitamins except vitamins D and B_{12}. But if the animal is grazing, it is getting sunshine and vitamin D, and ruminants can synthesize B_{12}. Rapidly growing grass is also rich in protein.

However, because of the high water content of succulent green feed (and roots, too), these are low in minerals. The lack of minerals, combined with the high water content (an animal could drown before it got enough nutrients from really lush grass), means that such forage does not constitute an adequate diet by itself. And lush forages can cause bloat.

This is not to say that green feed is not desirable, but only that it must be understood and properly managed.

Soiling in Confinement. There are two ways to use green forages. One is the familiar pasture, where the animals "harvest" their own feed. The other is *soiling*, where the caretaker does the harvesting and brings the feed to the animals. In this case, the goats are usually in *confinement;* that is, they are restricted to a loafing barn and a relatively small, and therefore mostly unvegetated, exercise yard.

Confinement feeding has been popular among people with a few goats for many reasons. Often land is limited. Goats are notoriously difficult to fence, and fencing large areas can be expensive. Unlike cattle or sheep, goats dislike grazing in inclement weather. Goats, more than cattle, are easy prey for stray dogs and other predators. Owners of a few goats are less likely than large farmers to be around during the day to keep an eye on things, or to have the time to manage rotational grazing. Animals on pasture waste a lot of feed by trampling and selective grazing. Pasturing allows the caretaker less control over what the animal eats, including toxic plants and those that can affect the flavor of milk. And because most home dairies with just a few animals (and outside jobs) purchase their hay and grain, feeding in confinement makes sense.

Farmers with sufficient acreage and equipment usually feed *green chop* — crops such as alfalfa or oats that are cut and chopped with tractor-drawn equipment and blown into special wagons. Some are feeders on wheels, while others require augering the feed out of the wagon and into feed bunks. Even if you have just a few goats, you can cut green feed with a scythe, string trimmer, or similar tool, or even a scissors or butcher knife, and carry it to the animals by the armload or in a barrow or cart.

Consider planting forage crops in your garden. Kale, chard, carrots, and others are popular, but oats, rye, and alfalfa grow well, too.

Grass Clippings

Goats are not lawn mowers and lawns are not pastures. And the better your lawn — from a homeowner's perspective, that is — the worse it is for goats. Goats like variety and dislike eating off the ground, perhaps as an instinctive defense against parasite infestation. Lawns strike out on both counts.

Lawn grasses have been developed to withstand foot traffic, frequent mowing, and for color, among other things, not for animal nutrition. They are high in moisture and very low in fiber and nutrients. They tend to pack tightly (both in the feeder and in the rumen) and they heat up and mold quickly. They take up rumen space that could better be used by more nutritious feeds. And because of their poor nutritive quality, you'd have to feed more expensive concentrates to provide a balanced diet.

If you have a "beautiful" lawn, it's probably because you applied nitrogen fertilizer, which can mean toxic levels of nitrates in the grass. And naturally, if herbicides have been used, that grass is totally off-limits to goats.

On the other hand, some of us don't have or even want picture-perfect lawns like that. Ours are a combination of grasses and weeds, perhaps including clovers and chicory, and of course, plenty of dandelions. We never fertilize the lawn and certainly never spray it.

In this case, one way to utilize that grass is to mow it without a grass catcher. Let it lie until it's dry. This might take a few hours on a hot dry day, or it might take a day or two. This is basically *lawn hay*. It can be fed to the goats in limited quantities, or if it's dry enough, stored in plastic bags or a hay mow. If you choose to feed fresh grass clippings, provide only as much as the goats will clean up in perhaps half an hour. Remove the leftovers before they heat up. And provide other feeds as well to lessen the packing effect of the grass clippings in the rumen. (See also Silage, page 90.)

Pasture

Pasturing goats presents both opportunities and problems. It seems like an obvious way to reduce feed costs — the major expense for the home dairy — especially if you have suitable land available. But before you take this economy for granted, check into the cost of fencing! (*Note:* One acre of land will require at least 825 feet of fencing, and more if it's not square.)

Also take an inventory of what's growing in your proposed pasture. Goats prefer variety, and woody plants rate high with them. Watch for poisonous plants (see page 86).

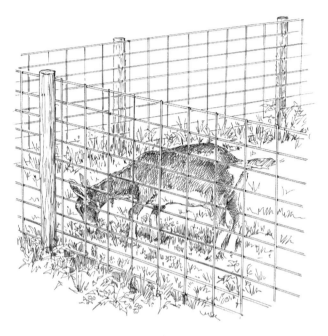

A well-fenced area with a variety of plants provides a fine pasture that makes raising goats easy and economical.

In most cases, animals on pasture trample and otherwise waste more than they eat. And goats are picky, taking a bite here, a nibble there. With their almost prehensile lips, they can select exactly what *they* want, which is not necessarily what *you* want for them. They don't graze down to the ground as sheep and cattle do, making it difficult to know just how much nutrition a goat is getting from browse. Laboratory tests of plants determine the food value of the entire plant, but goats generally select just the part they want.

Good Pasture Management. Good pasture management involves providing the best pasture in each season with a stocking rate that is compatible with good renewal of the vegetation and the best sustainability of forages and browse.

The seasonal factor is important, because different plants grow best under different conditions. Renewability and sustainability are important because left on their own, without rotation, goats (and sheep and cattle) will eat the plants they favor, perhaps killing them out. The less desirable plants will take over.

Seasonality also involves nutrition. A dramatic example can be seen in the lush pastures of spring becoming sparse and brown after a summer's hot and dry spell. On poor pasture a goat will not produce milk, and her health might even be endangered without supplemental feed.

Ideal pastures are soil-tested and properly fertilized; planted to specific desirable species; and managed to avoid overgrazing and to encourage overall productivity, preferably including rotational grazing.

Rotational Grazing. The simplest way to utilize pastureland is to fence it in, turn in the goats, and let 'em at it. But, much as most goat owners cherish simplicity, it isn't always the *best* way.

Any grazing or browsing animal introduced to a pasture will select their favorites, first. This isn't quite like a child who, given free rein at the dinner table, starts with the dessert: it's worse. The problem is damage to the pasture. The preferred parts of the preferred plants are eaten with relish, usually faster than they can regrow. Meanwhile, other plants that might be just as nutritious are left to grow old and coarse. The most-favored plants can be eradicated by overgrazing, while the others become useless.

With rotational grazing, the pasture is divided into paddocks, or smaller pastures. How many depends on several factors, including how much time and money you want to spend on fencing. Some people use only a few paddocks, while others have twenty paddocks or more.

The animals are allowed into one section, where they first seek out their favorite foods. But because there aren't as many of those in the smaller space, less-favored plants are also consumed.

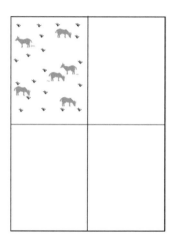

Rotational (or intensive) grazing involves moving animals from one paddock, or small pasture area, to another as the forage plants are consumed.

Then after a few days — or more, or less, depending on the degree to which the plants have been eaten — the livestock are moved to a new section. The feeding process starts all over again, while the previous pasture gets a chance to regrow.

Movable fencing makes this much more practical than it would be with permanent fencing, and the additional forage produced on the same amount of land can make such a system worthwhile. There are other benefits, such as increasing the sustainability of the land. One drawback, in addition to the need for additional fencing, is providing drinking water in each section. A good solution for goats is to design the fence layout so the animals can return to the barn to drink, or if it rains.

Controlled, rotational grazing provides many benefits, and if it interests you, you'll want to learn more about it. *Grass Farmer* is a publication devoted to this system.

Weeds and Poisonous Plants. However, a variety of weeds in a pasture can be both attractive and beneficial to goats. Some common ones include chicory, daisies, dandelions, nettles, plantain, thistles and yarrow. And some plants humans try to avoid are safe for goats, including nettle and poison ivy. Goats also love some serious pest weeds, such as multiflora rose, which are not mentioned in the scientific literature as having caused any problems.

Sorrel and dock are often considered valuable feed weeds, but these are well-known oxalate accumulators, which can be toxic. Spinach has been associated with oxalate toxicity of goats in Australia, as has amaranth (pigweed) in Mexico. Grazing goats are unlikely to consume enough of these to cause problems, but it's not a good idea to cut armloads for goats in confinement. (Oxalate toxicity also results from chemicals such as ethylene glycol — automotive antifreeze.)

Poisonous plants aren't a common problem with goats. You should certainly be aware of what's growing in your pastures and their possible effects on goats. Arm yourself with a good plant identification book for your region, and ask your county extension agent what problem plants you should be watching for. Goats' browsing habit of taking a bite here and a nip there tends to protect them from consuming too much of most dangerous plants. But be on the lookout for wilted wild cherry, oak leaves, rhubarb leaves, milkweed, hemlock, mountain laurel, and locoweed, all of which are poisonous to goats. Problems with bracken fern, which is toxic to cattle and sheep, are very rare in goats.

Good Weeds for Goats

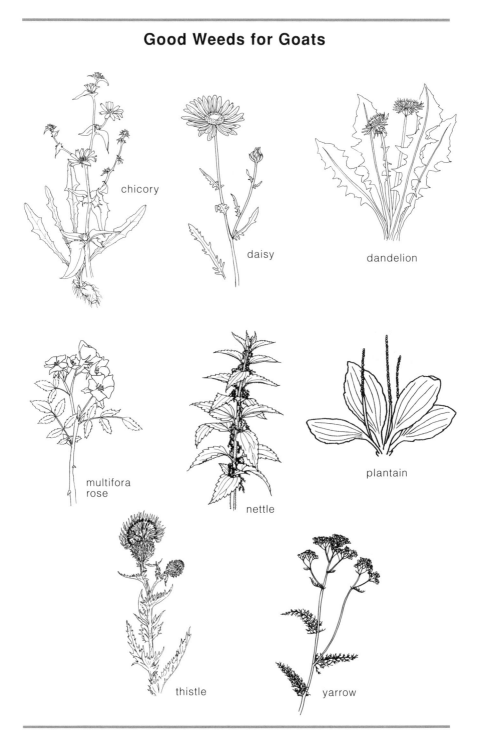

chicory

daisy

dandelion

multifora rose

nettle

plantain

thistle

yarrow

Bad Weeds for Goats

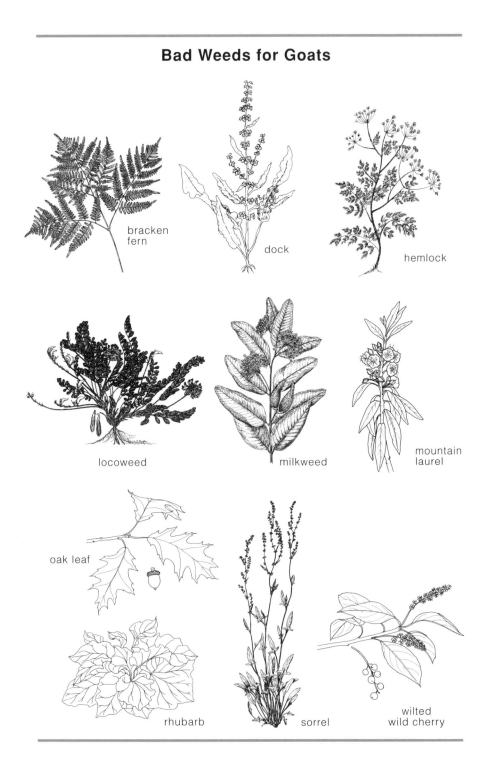

bracken fern

dock

hemlock

locoweed

milkweed

mountain laurel

oak leaf

rhubarb

sorrel

wilted wild cherry

Nitrate Poisoning

Goats convert nitrates to nitrites, as do plants. Nitrate poisoning is caused not by nitrates, but by their overaccumulation in the goat's system.

Some plants normally considered good feed can undergo chemical changes due to weather. Sudan grass, Johnson grass, pigweed, lamb's-quarters, alfalfa, corn, and oats can accumulate toxic amounts of nitrates if they undergo rapid growth after a dry spell. Low temperatures, decreased light, and, of course, heavily fertilized soils can also adversely affect these plants, causing nitrate accumulation.

Nitrate poisoning can also be caused by water contaminated with animal wastes or run-off from fertilized fields, or by eating fertilizers. Several cases of nitrate poisoning in goats have been blamed on water provided in galvanized containers.

Johnson grass   amaranth (pigweed)   alfalfa

Silage

Occasionally a *Countryside* reader asks about feeding silage to goats. Yes, they will eat it; and yes, you can make it in plastic garbage bags, even from lawn clippings. But it's much more complex than that, and very few goat owners bother because for most, it's more trouble than it's worth. But it doesn't hurt to know the basics so you can judge for yourself.

A silo is a huge pickling vat. Think of a sauerkraut crock 12 feet or more in diameter and often 80 feet high or more. Chopped corn — the whole plant, cobs, stalks, and all *(corn silage)* — or chopped hay *(haylage)* is blown into the top of the silo through a pipe, much like insulation is blown into a house. Proper chopping is essential to assure good packing, which eliminates air. It is firmly packed and left to ferment, just like kraut, for about 21 days. Note that the moisture content is of extreme importance. If it's too dry, it won't ferment; too wet, and it will rot.

Silage, or haylage, is a standard feed on most cow dairy farms because it's nutritious, easy to handle, and very cheap compared to hay and grain. But these attributes are hard to capture in a goat dairy, especially a small one. One of the primary reasons is spoilage. This occurs when silage is exposed to air, so a certain amount must be used every day. With a conventional silo, this is easy for cow herds, but almost impossible for most goat herds.

Farmers like corn silage because it utilizes the entire plant, thus producing more feed (and milk) per acre. In northern areas this use can salvage corn that doesn't mature enough to harvest as grain.

Garbage Bag Silage

Despite its problems, silage appeals to many small landholders, especially when they hear that it can be made from lawn clippings, in garbage bags! The method differs from "real" silage only in scale. The requirements for moisture levels and air exclusion are the same. And remember, if grass has been treated with chemical fertilizers or pesticides, it's not safe for goats. Here's how it's done.

1. Mow the grass and let it dry, but not too much. Depending on the weather, and humidity, and the moisture content of the grass, drying might take only a few hours, or it might take a day or more. Remember: Too dry and it won't ferment, too wet and it will rot. (Good haylage has about 30 to 35 percent dry matter, but more is recommended for bag silage.)
2. Rake up the grass, and eliminate any twigs or stems that might puncture the bag. Pack the clippings tightly into the strongest plastic bags you can find. Sit on the bag to force out as much air as you possibly can. Tie the top tightly, using baling twine not a twist tie.
3. Store the bags out of the sun for 3 weeks. As long as the bags don't get punctured or opened, the silage will keep for many months.

Note: When you open the bag, there might be a thin layer of white mold on the top. This is normal, since it's almost impossible to exclude all the air, which activates the bacteria in the grass. Put the moldy stuff in the compost bin. If the whole bag of grass is a slimy, stinking mess,

Haylage doesn't require the dry weather needed to make baled hay, and can even salvage forages that were intended for hay but got caught by wet weather. It can also be made early or late in the season, when good drying weather is often scarce. Haylage can be made from alfalfa or other legumes; forages such as fescue and other grasses; or cereal grains in the "boot" stage, including oats and rye.

In one Midwestern study, alfalfa silage had 4 percent more protein than alfalfa of the same quality put up as hay, because of reduced leaf loss. In addition, the dry matter harvest was 27 percent higher.

Silage is much more mechanized than either grain or baled hay. There are no bales to store in the haymow and no bales to throw down at feeding time. No grain dryers, no grinding and mixing. An auger on a silo

it's *all* compost. Either the grass went into the bag too wet, or there was air in the bag. Good grass silage is green and has a fermented but pleasant aroma. If you're really short of feed, or money to buy it, bag silage might be worth the effort. Under any circumstances it could be fun to try it, just as an experiment. But the average backyard goat dairy will be time and money ahead by feeding purchased hay and a prepared goat grain ration.

Use a bag, once opened, within 2 to 3 days. Just a few goats won't eat this much, especially at first when you'll naturally want to limit feeding it to avoid digestive upsets. Always introduce any new feed slowly, over a week or more.

"Garbage bag silage" can be made from lawn clippings, if you don't use pesticides on your lawn. Eliminate as much air as possible from the bag before closing it tightly, and be sure the plastic won't be punctured. Air will cause the clippings to mold and rot.

unloader scrapes off the top layer and shoots it down to a waiting cart (often electric, these days) that delivers it to the animals. In some operations it goes directly to the feed bunks.

Feed Value. As for feed value, the fermentation in the absence of air causes the sugars to break down, which is like predigesting the feed for the animals. The acidity increases, which accounts for its keeping qualities, but this keeping quality is lost in the presence of air. Oxygen activates the bacteria and the silage heats up and molds within 2 to 3 days. That, and the cost of equipment, is the bad news for goat milkers.

If enough material can be removed from the top of the silage every day this is no problem. But in even a small silo, this amounts to far more than most goat herds can consume in a day. With just a few animals, forget it.

There are other problems. We know of one case where a cow dairyman also had a number of goats. The cows supposedly took care of the spoilage problem, so the goats got silage too. Some of them got sick. The problem was identified as listeriosis, traced to the silage. It didn't bother the cows.

Silage is a relatively recent agricultural development. Its original purpose was to replace labor-intensive root crops such as carrots, mangel beets, and turnips with more "industrialized" capital-intensive feeds. As farms got bigger and more mechanized, research concentrated on silage rather than root crops. Today, silage is common, while almost no one grows roots for livestock feed, though some homesteaders do.

Hay

Understanding hay is an important part of raising goats. But many people who are new to goats have no background with farm animals and no experience with hay. Some don't even know the difference between hay and straw. *Hay* refers to grasses or legumes cut at an early stage of growth and sun-dried. Hay should be more-or-less bright green and is used as feed for equines and for ruminants such as goats. *Straw* is the dried leaves and stems of grasses grown for grain, such as oats, wheat, and barley, after the grain has been removed by combining. Straw is a golden color and is used as bedding, seldom as feed.

Not all hay is created equal. The first major division is between grass or carbonaceous hay, and legume hay. Grass doesn't refer to lawns! *Grass hay* might be timothy, Johnson grass, orchard grass, or others grown specifically as animal feed. *Legume hay* comes from alfalfa or the clovers.

The carbonaceous hays have less protein and less calcium than the legumes, and these deficiencies must be made up in the concentrate or grain ration.

Other hay plants can include barley (cut when the seed heads are in the immature, or "boot," stage), bird's-foot trefoil, Bermuda grass, lespedeza, marsh or prairie grasses, oat or wheat grasses, soybeans, or combinations of these. For the benefit of inexperienced farmers, it will be well to point out again that hay is made by cutting green growing plants and drying, or *curing*, them in the sun. Wheat, barley, oats, and soybeans, for example, can be cut when young, for hay. If allowed to mature, the nutriment goes into the grain and the stems and leaves become yellow and have little food value, and the plant that could have been hay becomes straw.

Hay can also be first, second, or third crop or cutting (and more in some climates). The *first crop* generally has coarse stems and less total digestible nutrients when compared with later cuttings. The stage of cutting legumes is crucial. Alfalfa and clover should just be starting to blossom. After that the nutritive value decreases rapidly. Some hayfields include weeds of various kinds, which must be considered when trying to determine the feed and dollar value of the hay.

Hay Quality. But all of this can be overshadowed by how the hay is *cured* in the field. If it rains while the hay is drying, the feed value can be diminished, or even totally destroyed. If the hay is baled while too wet (either from rain or inadequate curing) mold is a problem; too dry, and the leaves containing most of the nutrition are likely to be shattered and lost in the baling process. Some hay is conditioned by passing through rollers that crush the stems, allowing for more rapid and uniform drying of stems and leaves.

The importance of quality hay can be illustrated by the fact that good alfalfa can have as much as 40 milligrams of carotene per pound while alfalfa that is bleached and otherwise of poor quality can have as little as 4 milligrams per pound. Poor hay may be difficult to distinguish from some straw. Straw contains much fiber, especially lignin, and is used as stock feed only in emergencies or under unusual circumstances.

Alfalfa or clover hay is considered the ideal for goats because of the high protein content. Good alfalfa has about 13 percent protein; timothy and brome are usually closer to 5 percent. Alfalfa and clover are also rich in calcium, the most important mineral.

Experienced producers and buyers can tell a great deal about hay by its color, texture, aroma, and general appearance. But your county extension agent can also tell you where and how to have hay tested so you know exactly what you're getting. *Forage testing* can help you determine what kind of grain ration to supplement hay with, for both performance and economy. If your hay is very good you might use a grain mix with less protein, which can save money. If the hay has less protein than the average, you'll want to adjust the concentrate to compensate for that. Just as important, hay testing can help determine a fair price.

Caution

Don't feed white clover; it's toxic.

Hay Prices. Hay prices vary considerably from place to place and year to year. It's not unheard of for hay in the Midwest to go for $1 a bale and $10 a bale in California. (But bales are bigger in California.) A mature goat will require anywhere from 3 to 10 pounds of hay per day, depending on type, quality, waste, and other factors.

Weather is a big determinant. If the rains are timely and sufficient, farmers have plenty of hay and the price goes down. In times of drought, however, farmers might not have enough to feed their own animals, and the price goes up. In some areas alfalfa winterkill, the alternate thawing and freezing that force the crowns out of the ground, can cause prices to shoot up.

The most elusive factor for most goat owners is quality. Pure alfalfa, properly cut in the early bud stage, conditioned, and properly cured, generally costs more than grass hay. Fine-stemmed hay is worth more than coarse-stemmed hay. The first cutting, or first crop, is usually cheaper than second or third, but this distinction may be meaningless in an area where in a given year, there is no second or third cutting.

But then, you might not need or even want the very best-quality hay. Dry does and bucks might get different hay than the milkers, and coarse hay provides more body heat than fine during the cold days of winter.

Grains: The Concentrate Ration

While roughages are the most important part of the diet of a ruminant, they alone don't provide all of the needed vitamins and minerals, nor do they provide sufficient energy. Alfalfa hay has about 40 therms per

100 pounds; corn and barley each have twice that. Especially if you feed carbonaceous hays, which have only 5 percent protein, you must provide additional protein and calcium. Hays do not provide sufficient phosphorus. These missing elements are provided in the concentrate ration.

The concentrate ration is often called the *grain ration*, but this term can be misleading. Here's why.

For lactating animals the protein content of the concentrate ration should be about 16 percent if the roughage is a good legume. With less protein in hay, more must be added to the concentrate. For dry does, 12 percent protein is sufficient. Corn has about 9 percent protein, and only 6.7 percent digestible protein. Oats has about 13 percent protein and 9.4 percent digestible protein. Therefore, a mixture of equal parts of corn and oats would contain 11 percent protein or about 8 percent digestible protein. Clearly, these grains alone will not meet the demands of the growing or milking animal. Therefore, protein supplements in the form of soybean-oil meal (sometimes listed on feed tags as SOM), or linseed, or cottonseed-oil meal must be added. Many goat owners try to avoid cottonseed-oil meal because of the massive amounts of pesticides used on that crop.

Milking animals also require more salt than is needed for animals on maintenance rations. It is usually added at the rate of 1 pound per 100 pounds of feed.

Because of the need for bulk in the diet of a ruminant, a concentrate ration should not weigh more than 1 pound per quart. Bran is most commonly used for bulk. (Beet pulp is sometimes used for does, but extended feeding of beet pulp to bucks can cause urinary calculi.) The weight of grain varies with quality, which is often determined by the weather during the growing season, but this chart shows some normal averages.

Weights of Some Common Goat Feeds

INGREDIENT	WEIGHT IN POUNDS PER QUART
Barley, whole	1.5
Buckwheat, whole	1.4
Corn, whole	1.7
Linseed meal	0.9
Molasses	3.0
Oats	1.0
Soybeans	1.8
Sunflower seeds	1.5
Wheat, whole	1.9
Wheat bran	0.5

Molasses: An Important Extra

Finally, since goats generally shun dusty ground feed such as that normally fed to cows, the grains should be crimped or cracked or even whole, rather than ground to flour. Cows do not digest whole grains well. Whole corn goes in one end and out the other. Goats seem to have better powers of digestion.

This is fine for the goat owner who wants to mix a ration rather than buy bagged feed, because it eliminates the bother and expense of grinding. But it also means the fine ingredients — salt, bran, oil meal, and minerals — can't be mixed into the grain. They sift to the bottom and the goats won't consume them. To overcome this, most goat feeds contain cane molasses. In addition to binding the ingredients, molasses makes the feed less dusty, it's an important source of iron and other important minerals, it increases the palatability of the feed, and does fed ample molasses during gestation are less likely to encounter ketosis (see chapter 8). Molasses contains about 3 percent protein, but none of it is digestible.

There is some evidence, at least in dairy cows, that excess molasses intereferes with the digestibility of other feeds. The digestive processes attack the more easily assimilated sugars in molasses to the detriment of other feedstuffs. This is one reason some authorities advise against giving goats horse feeds, which are high in molasses. (Many also contain copper in amounts toxic to goats.) Feeds with more than 5 to 6 percent molasses should be avoided. Even so, molasses is an important feed for goats.

Creating a Balanced Ration

At last we're ready to formulate a ration. The main tool we'll use will be a list of the protein content of the common goat feeds (see pages 98–99). The idea is to combine the various ingredients you have available in such a way that the combination will contain the desired amount of protein or, more accurately, digestible protein. But since protein is only one element of feed value, we must also keep in mind the minerals, vitamins, fiber, and palatability.

Just as importantly, any given ration depends on locally available ingredients and their comparative prices, and the suggested rations almost invariably have to be adjusted. Unless the feeder knows what to look for, the carefully formulated suggestions will be thrown out of balance by indiscriminate substitutions.

Likewise, even the person who feeds commercial rations can destroy the balance by haphazardly adding "treats" or by making use of available grains in addition to the commercial feed. You can no more prepare a balanced diet by adding a handful of this to a scoop of that than you could expect to bake a cake by using the same method.

To complicate matters, the feed value of hay and grain varies from place to place and year to year, being affected by soil, climate, and other factors. For example grains grown in the Pacific Coast are lower in protein than those grown elsewhere; also, the nutrients in hay harvested at the proper stage of development and well cured will differ dramatically from hay that is cut too late and leached or spoiled by improper curing. (*Reminder*: Moldy hay should *never* be fed.)

Is It Worth the Trouble?

Is it worth going through all this, when it's so easy to buy scientifically formulated feed in convenient 50-pound bags? For most people, the answer is a resounding "No!" Initially, many people are shocked by the cost of commercially prepared bagged feeds. After all, they reason, if you can buy 100 pounds of corn for $5 or $6 dollars, why pay $15 or more for 100 pounds of feed mixed specifically for goats? Their question is answered when they stop to consider the hassle and expense of buying, storing, handling — and mixing — the individual ingredients.

On the other hand, some people want "organic" feeds that have been grown without chemical fertilizers, herbicides, or other pesticides, and processed without antibiotics, preservatives, or medications. Today, we can add genetically modified plants to this list. You might not be able to find organic commercial feeds, but if you have a source of organically grown grains, you can mix your own.

Then there are people who already grow hay and grain, or who will, when they get some goats. Naturally, they'd rather feed that to the goats than buy a commercial mix.

These rations are provided for those dedicated do-it-yourselfers. But studying them will help you get a better idea of what a goat diet should be like, even if you buy your feed ready-mixed.

Average Composition of Selected Goat Feeds

FEED	CRUDE PROTEIN	DIGESTIBLE PROTEIN	FAT	FIBER
Alfalfa hay	15.3%	10.9	1.9	28.6
Bermuda grass	7.1	3.6	1.8	25.9
Birdsfoot trefoil	14.2	9.8	2.1	27.0
Brome	10.4	5.3	2.1	28.2
Red clover	12.0	7.2	2.5	27.1
Mixed grass	7.0	3.5	2.5	30.9
Johnson grass	6.5	2.9	2.0	30.5
Soybean, early bloom	16.7	12.0	3.3	20.6
Timothy, early bloom	7.6	4.2	2.3	30.1
SUCCULENTS				
Green alfalfa, early bloom	4.6	3.6	0.7	5.8
Bermuda grass, pasture	2.8	2.0	0.5	6.4
Cabbage	1.4	1.1	0.2	0.9
Carrot roots	1.2	0.9	0.2	1.1
Kale	2.4	1.9	0.5	1.6
Kohlrabi	2.0	1.5	0.1	1.3
Mangel beets	1.3	0.9	0.1	0.8
Parsnips	1.7	1.2	0.4	1.3
Potatoes	2.2	1.3	0.1	0.4
Pumpkins (with seeds)	1.0	1.3	1.0	1.6
Rutabagas	1.3	1.0	0.2	1.4
Sunflowers (entire plant)	1.4	0.8	0.7	5.2
Tomatoes (fruit)	0.9	0.6	0.4	0.6
Turnips	1.3	0.9	0.2	1.1
Barley	12.7	10.0	1.9	5.4
Steamed bone meal	7.5	—	1.2	1.5
Buckwheat	10.3	7.4	2.3	10.7
#2 dent corn	8.7	6.7	3.9	2.0
Linseed meal	35.1	30.5	4.5	9.0
Cane molasses	3.0	—	—	—
Oats	12.0	9.4	4.6	11.0
Field peas	23.4	20.1	1.2	6.1
Pumpkin seed	17.6	14.8	20.6	10.8
Rye	12.6	10.0	1.7	2.4
Soybeans	37.9	33.7	18.0	5.0
Sunflower seed, with hulls	16.8	13.9	25.9	29.0
Wheat (average)	13.2	11.1	1.9	2.6
Wheat bran	16.4	13.3	4.5	10.0

Nitrogen-Free	Mineral Matter	Calcium	Phosphorus Extract
36.7	8.0	1.47	0.24
48.7	7.0	0.37	0.19
41.9	6.0	1.60	0.20
39.9	8.2	0.42	0.19
40.3	6.4	1.28	0.20
43.1	6.5	0.48	0.21
43.7	7.5	0.87	0.26
37.8	9.6	1.29	0.34
44.3	4.7	0.41	0.21
9.3	2.1	0.53	0.07
12.2	3.1	0.14	0.05
4.4	0.7	0.05	0.03
8.2	1.2	0.05	0.04
5.5	1.8	0.19	0.06
4.3	1.3	0.08	0.07
6.0	1.0	0.02	0.02
11.9	1.3	0.06	0.08
17.4	1.1	0.01	0.05
5.2	0.9	—	0.04
7.2	1.0	0.05	0.03
7.9	1.7	0.29	0.04
3.3	0.5	0.01	0.03
5.8	0.9	0.06	0.02
66.6	2.8	0.06	0.40
3.2	82.1	30.14	14.53
62.8	1.9	0.09	0.31
69.2	1.2	0.02	0.27
36.7	5.7	0.41	0.85
61.7	8.6	0.66	0.08
58.6	4.0	0.09	0.33
57.0	3.0	0.17	0.50
4.1	1.9	—	—
70.9	1.9	0.10	0.33
24.5	4.6	0.25	0.59
18.8	3.1	0.17	0.52
69.9	1.9	0.04	0.39
53.1	6.1	0.13	1.29

Sample Ration Formulas

These rations, followed more or less faithfully, could be expected to produce good results. There will be minor variations because the feed value of grains depends in part on variety, weather, and the fertility of the soil that produced them. Most grains grown in the Pacific Northwest, for example, are lower in protein than the same grains grown elsewhere; and old-fashioned open-pollinated corn has more protein than today's hybrids.

Sample Rations

FEED	POUNDS
For a milking doe fed alfalfa hay (for a total of 12.6% digestible protein)	
Corn	31
Oats	25
Wheat bran	11
Linseed-oil meal	22
Cane molasses	10
Salt	1
or	
Barley	40
Oats	28
Wheat bran	10
Soybean-oil meal	11
Cane molasses	10
Salt	1
For a dry doe or buck (for a total of 9.8% digestible protein)	
Corn	58
Oats	25
Wheat bran	11
Soybean-oil meal	5
Salt	1

FEED	POUNDS
For a milking doe fed non-legume hay (for a total of 21.2% digestible protein)	
Corn	11
Oats	10
Wheat bran	10
Corn-gluten feed	30
Soybean-oil meal	24
Cane molasses	10
Salt	1
or	
Barley	25
Oats	20
Wheat bran	10
Soybean-oil meal	25
Linseed-oil meal	15
Salt	1
For a dry doe or buck (for a total of 10.1% digestible protein)	
Barley or wheat	51.5
Oats	35
Wheat bran	12.5
Salt	1

However, there are more serious considerations than these. One is that certain ingredients might not be available in your locale, or others may be more common and therefore less expensive than those listed. Grains can be substituted for one another by using the chart showing protein contents on pages 98–99.

Determining Pounds of Protein

You can determine the weight of protein in a given feed ingredient by multiplying the number of pounds of the ingredient by its percent of digestible protein. If you work in batches of 100 pounds, to figure the percent of protein in the whole ration, merely move the decimal point two places to the left to get the percent of protein in a ration.

As an example, let's look at a small homestead farm that produces its own grain. The previous year's corn crop was almost a total failure due to a wet spring, summer drought, and early frost. But other grains were available. Here's what the milking does were fed:

FEED	WEIGHT (LBS)	% CRUDE PROTEIN	% DIGESTIBLE PROTEIN	PROTEIN (LBS)
Soybeans	20	37.9	33.7	6.74
Barley	19	12.7	10.0	1.9
Oats	20	13	9.4	1.88
Buckwheat	5	10	7.4	0.37
Wheat bran	5	16.4	13.3	0.66
Corn	10	9	6.7	0.67
Linseed meal	10	34	30.6	3.06
Molasses	10	3	0	0
Salt	1	0	0	
TOTAL	100			15.28

Divide the pounds of protein (15.28) by the total weight of the ration (100 pounds). This feed has 15.28 percent protein.

It should be noted that some rations you will find elsewhere work with crude protein rather than digestible protein. Since there are no digestibility trials on goats, both methods have flaws. It may be easier to obtain figures on crude protein for locally grown feeds from your county extension office, in which case all the ingredients should be calculated on the basis of crude protein.

A Few Special Considerations

This leads us to another — perhaps the most important — reason why every goat owner should have at least a basic knowledge of feed formulations. Goat keepers are notorious for dishing out treats or making use of "waste." These are both admirable pursuits, but they can cause trouble.

Watch the Balance. Assume a goat receives 1 pound of a commercial 16 percent (crude) mixture. Maybe it costs the owner $16 per cwt, and he can get corn for half that, or he grew a little corn for the chickens and has some extra. Or the goat just seems to "like" corn! So he decides to give the goat ½ pound of the regular ration and ½ pound corn. So out of 100 pounds of feed, the goat gets 50 pounds of goat feed, which amount to 8 pounds of protein, and 50 pounds of corn, which amounts to 4.5 pounds of protein. The total protein per 100 pounds of feed is 12.5 pounds or 12.5 percent. So the goat's protein intake has been reduced from 16 percent in the original feed mixture to 12.5 percent. That might be enough for the goat to maintain her own body, but not to produce kids and milk.

The same thing happens when the animal is given garden "waste" or trimmings. Such fodder replaces roughage, not grain, but even then it can cause unbalancing of the diet because elements of hay, for instance, will be missing from most of the garden produce.

This is not to say that rations can't be manipulated or that the goat breeder shouldn't make use of what's available or cheap. Just do it with a certain amount of knowledge and discretion.

With this principle firmly in mind, let's examine some of the common feeds small farmers have available and show an interest in.

Soybeans. Soybeans deserve special mention because many people look at the price of the oil meals and wonder why the beans can't be fed whole. They can, with certain restrictions.

Soybeans contain what is called an *antitrypsin factor*. Trypsin is an enzyme in the pancreatic juice that helps produce more thorough decom-

Feed-Cost Guide

To be profitable, total feed costs should *not* exceed one-half the value of the milk produced.

position of protein substances. The antitrypsin factor doesn't let the trypsin do its job, which means the extra protein in the soybeans is lost, not digested. The antitrypsin factor can be destroyed by cooking and isn't present in soybean-oil meal.

However, rumen organisms apparently inactivate the antitrypsin factor when raw soybeans are fed in small amounts. Current recommendations for dairy cattle are that the ration not contain more than 20 percent raw soybeans. The same seems to work for goats. There is one major exception: do *not* feed raw soybeans if your feed contains urea! The result will probably be a dead goat. Here's why.

Urea. Urea is a nonprotein substance that can be converted to protein by ruminants. Urea is not recommended for goats, but many dairy feeds for cows contain it. So does liquid protein supplement, or LPS, which some feed dealers will try to sell you when you ask for molasses. (Such dealers are not being dishonest. LPS is okay for cows, but not goats, and often they simply don't know as much as you do about our caprine friends.) Some people feed urea to goats because it's less expensive than the oil-meal protein supplements, but many goat keepers have reported breeding problems with animals fed urea. Toxicity can result from improperly mixed feed or when urea is fed along with a high-fiber diet that lacks readily digestible carbohydrate.

Plant a Goat Garden

Most goats are raised on small farms or homesteads where grain and hay are not produced. Such places can still grow a great deal of goat feed if the basic principles of feeding are followed.

"Grow milk in your garden" by planting sunflowers (the seeds are high in protein and the goats will eat the entire plants), mangel beets, Jerusalem artichokes, pumpkins, comfrey, carrots, kale, and turnips, among others. In addition, you can utilize such "waste" as cull carrots and apples and sweet corn husks and stalks in the goat yard. Treat them like pasture or silage, using them to replace part of the grain ration, but not all of it. Feed at least 1 pound of concentrates per head per day to milking animals.

Gathering Weeds for Your Goats. Many people with more time than money and a keen interest in nutrition are also avid collectors of weeds for their animals. For example, dandelion greens are extremely rich in vitamin A, and nettles are high in vitamins A and C. Goats relish these and other common weeds. It's just about impossible to imagine a real farmer on his knees gathering dandelion greens for his livestock, but many a goat farmer does, and reaps healthier animals and lower feed bills.

Goats seem to enjoy variety more than most domestic animals, and no one plant has everything any animal needs for nutrition. Many goatkeepers provide food from as many different plant sources as possible to enhance the possibility that their animals are getting the nutrition they need, naturally, without synthetic additives. They like grain mixtures of at least five or six different ingredients.

Obviously, this isn't as efficient as modern agricultural methods. Farmers know that alfalfa is rich in protein and calcium, both important to dairy animals. A great deal of feed can be harvested from 1 acre of alfalfa, and alfalfa hay has become the norm. There are even herbicides to kill weeds in alfalfa to keep stands pure.

But almost any weed in your garden has more cobalt than alfalfa. And cobalt is required by ruminants to provide the bacteria in the digestive tract with the raw material from which to synthesize vitamin B_{12}. Some, if not all, internal parasites rob their hosts of this vitamin.

Alfalfa and clover have little cobalt because lime in the soil depresses the uptake of this mineral, and lime is necessary for the growth and the calcium content of each. Agribusiness has found it more efficient to strive for high yields of alfalfa and then add the trace minerals to the concentrate ration. Homesteaders who don't mind gathering weeds can meet their animals' nutritional needs naturally and without the cash outlay required for commercial additives.

Organic farmers have known this for years, of course, but when their beliefs were confirmed by scientists in 1974 the idea was hailed as revolutionary. Researchers at the University of Minnesota compared the nutritive value and palatability of four grassy weeds and eight broadleaf weeds with alfalfa and oats as a feed for sheep, which have roughly the same requirements as goats.

Caution

Don't gather weeds from along roadsides where spraying is done.

Lamb's-quarters, ragweed, redroot pigweed, velvetleaf, and barnyard grass all were as digestible as alfalfa and more so than oat forage. All five weeds had more crude protein than oats and four had as much as alfalfa. Eight were as palatable as oat forage.

Goats Love to Trim the Trees

Tree leaves and bark can be rich sources of minerals, brought from deep within the earth by tree roots. As far as goats' diets go, tree trimmings fall into the category of weeds. Some are great for goats; a few are dangerous. Here's a sampling.

- ◆ **Pine boughs.** Goats love them, but there have been reports that pine needles have caused abortions, so caution is advised. Pine boughs are rich in vitamin C, although goats have no particular need for the vitamin, being able to manufacture it themselves.

- ◆ **Apple trees.** These are a great treat. By all means, feed the trimmings of apple trees when you prune your organic orchard.

- ◆ **Wild cherry.** Avoid altogether, since the wilted leaves are poisonous.

- ◆ **Oak leaves.** Avoid, too, as they're toxic for goats.

Comfrey. One particular plant deserves special attention, because so many people are interested in it and because it's controversial. That's comfrey, also known as boneset.

Several years ago there were a rash of statements from county extension agents and state departments of agriculture knocking comfrey. Some of their reasons for not growing it are practical — for large farmers, not homesteaders. And some of their information is just plain wrong.

It is true that a study conducted in Australia some years ago suggested that comfrey might be carcinogenic when fed excessively or over an extended period. That study didn't involve goats. And since then, goats have consumed tons of comfrey, with no problems showing up in the scientific veterinary literature.

Even aside from that, comfrey should be in every goat owner's garden for at least limited use. Many goat and rabbit raisers swear by comfrey as a feed, a tonic, and as medication for certain conditions such as scours. It is high in protein, ranking with alfalfa, although there is some question about the digestibility of the protein. But it is easier to grow and harvest than alfalfa, using hand methods. It is an attractive plant that even can be used for borders or other decorative applications: grow goat food in your front yard or flower bed! It has tremendous yields since it begins growing early in spring and grows back quickly after cutting. And it is a perennial. It can be dried for hay although that entails a lot of work because of the thick stems. It must be cured in small amounts on racks rather than left lying on the ground.

comfrey

Comfrey Salve

If you're into herbal medicine, you're aware that comfrey has valuable healing powers for humans. This applies to cuts and wounds on goats as well. For winter medicinal use, thoroughly mash four to five large, clean comfrey leaves. Mix the mashed leaves with ¼ cup of unhydrogenated oil (found in health food stores). Refrigerate. Apply to your goats' wounds as needed.

Minerals. No discussion of goat feeds would be complete without mentioning minerals. Most goat raisers supply loose minerals or mineralized salt blocks free choice, and also add dairy minerals to the feed.

While there is no sound research into the matter, there is some indication that this is unnecessary, expensive, and perhaps even dangerous. Too much of a good thing can be as bad as too little. Goats that are well-fed on plants and plant products from a variety of sources, grown on organically fertile soil, probably have little or no need for additional minerals.

There are certain exceptions. Plants grown in the goiter belt (from the Great Lakes westward) are low in iodine, and iodized salt will be good insurance. Certain areas of Florida, Maine, New Hampshire, Michigan, New York, Wisconsin, and western Canada have soils deficient in cobalt. Parts of Florida are deficient in calcium.

Selenium deficiency can cause white-muscle disease (see chapter 8). Most soils in the central and eastern United States (and some in other areas) are deficient in selenium. Injecting does with selenium 15 to 30 days before kidding and kids at 3 to 4 weeks of age will prevent this serious disease. There's a relationship between selenium and vitamin E; both are usually administered at the same time. On the other hand, some soils, and the feeds grown on them, are high enough in selenium to cause poisoning.

Phosphorus is a vital ingredient of the chief protein in the nuclei of all body cells. It is also part of other proteins, such as the casein of milk. Therefore, it is of extreme importance to growing animals that are producing bone and muscle; pregnant animals that must digest the nutrient needs for the growing fetus; and for lactating animals, which excrete great quantities of these minerals in their milk. Vitamin D is required to assimilate calcium and phosphorus. Also, the ratio of calcium to phosphorus is critical: it should be 1.5:1.0. When legume forages are fed, the goat might need more phosphorus in the concentrate ration to maintain the 1.5:1.0 balance; when carbonaceous forages are used, supplemental calcium might be required. Roughages, especially legumes, are high in calcium, and grains are high in phosphorus. If these crops are grown on soils rich in these minerals the well-fed goat is likely to get enough of them.

A goat lacking phosphorus will show a lack of appetite, it will fail to grow or will drop in milk production if in lactation, and it may acquire a depraved appetite such as eating dirt or grawing on bones or wood. (Many goats like to chew on wood even without a phosphorus deficiency.) In extreme cases stiffness of joints and fragile bones may result.

However, overfeeding calcium can be dangerous, too, especially for young animals. Lameness and bone problems can result later from excess calcium.

Iron is 0.01 to 0.03 percent of the body, and is vital for the role it plays in hemoglobin, which carries oxygen in the blood.

Copper requirements are about one-tenth those of iron, and in greater amounts copper is a deadly poison.

Nutritional anemia can result from lack of iron, copper, or cobalt. (This is different from pernicious anemia in humans.) But it's very rare.

Other trace minerals are potassium, magnesium, zinc, and sulfur.

Most goat owners today prefer furnishing loose minerals free choice, rather than in block form, and find that their goats consume varying amounts depending on their needs. In this case there is no need to add minerals to the feed too; as we have seen, it might not be a good idea.

Do Goats Know What They Need?

Some livestock raisers believe animals can select the minerals they need. Theta Torbert of Maine told *Dairy Goat Journal* that her goats use a lot of iodine during the third month of pregnancy, magnesium just before they go on spring pasture, sulfur in August and January when they are growing their fur coats, phosphorus during the last 2 months of pregnancy, and a lot of potassium year-round. They don't seem to use calcium, but they do get dicalcium phosphate in their grain. If you want to check this out for yourself, furnish the loose minerals in separate mineral feeders.

The Science and Art of Feeding

Obviously, feeding is a science and an art. Goats are not "hayburners" or mere machines to be fueled haphazardly. You wouldn't burn kerosene in a high-powered sports car, and you can't get the full potential from a goat fed improperly.

Remember these basic concepts:

- Feed your goats 1 pound of concentrate for maintenance and about 1 pound extra for each 2 pounds of milk produced, along with all the hay they will eat.
- The ration should come from as many different sources as possible, and fertile soil.

◆ Avoid sudden changes in feed, which result in overloading the rumen bacteria and microbes.

◆ Pay attention to protein levels as well as vitamin and mineral content of the plants and grains you feed.

◆ Treat each animal as an individual, for they have different needs according to age, condition, production, and personal quirks.

◆ Some will do better on less, others will want more. That's the art, or part of it: "The eye of the master fattens the livestock."

GROOMING

Goats require a minimum of care, but that doesn't mean they require *no* care. The goat owner will quickly learn how to disbud, tattoo, clip, and trim hooves.

Let's begin with the most important: hoof trimming.

Hoof Care

The horny outside layer of the goat's hoof grows much like your fingernails and, like nails, must be trimmed periodically. Gross neglect of this duty can cripple the goat. But it's a simple job, and for the small herd it won't take more than a few minutes every few months. This makes it hard to understand why there are so many goats whose feet look like pointy-toed elf boots, or even skis.

How Often?

How often you trim hooves depends on several factors. But at the very least, you should check hooves every 2 to 3 months. Sometimes hooves grow faster than at other times, and there are differences among animals. Goats living on soft spongy bedding will need more hoof attention than goats that clamber on rocks, which is how wild goats' hooves are worn down. One goat book claims that if a good-sized rock is placed in the goat pen, the animals will stand on it and keep their hooves worn down. It sounds logical, but everyone I know who has tried it says it doesn't work.

Natural Hoof Maintenance

In nature, goats' hooves are worn down when they climb on rocks. If goats live on bedding or soft ground, hoof trimming becomes the job of the caretaker. Increase the time between hoof trimmings by placing a dozen or so rough cinder blocks in the yard or pasture, with holes turned to the side, obviously. The goats will love to climb on them, and the abrasive surface of the blocks will help minimize your trimming duties.

What Tools?

There are several methods of trimming, requiring different tools. The simplest trimming tool is a good sharp jackknife. Some people prefer a linoleum or roofing knife, and others swear by the pruning shears, the same kind you use for your roses. However, I don't think anyone who has used an honest-to-goodness goat hoof trimmer would ever want to go back to the more primitive tools. (Some catalogs selling sheep equipment call them "foot rot shears," a rather ugly and misleading name for the same tool.)

For light trimming and finishing, many people like a Surform, a small woodworker's plane with blades much like a vegetable grater.

A stiff brush for cleaning and heavy gloves are also useful.

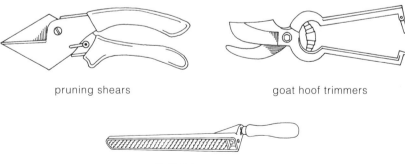

pruning shears   goat hoof trimmers

Surform

How to Trim Hooves

Unless you have docile goats or are rather robust yourself, hoof trimming is easier if you have a helper, or if your milking stand is of a type that allows you to lock the goat in and still have room to get at all four feet (see chapter 13 for more information on milking stands).

1. With the goat secured, stand ▶ **against her rear (tail to tail, as it were) and grasping one hind leg, lift it up between your legs.** (Some goats don't seem to mind such acrobatics; others will protest rather violently.) Keep a firm grip on the ankle, and be exceedingly careful with the knife or other tool so that in case she does kick, neither one of you will be injured. This is one reason many people prefer the shears, which is much safer than a knife. In either case, it's a good idea to wear heavy gloves.

2. With the point of the tool or ▶ a stiff brush, **clean out all the manure and dirt embedded in the hoof.** If the hoof has not been trimmed for some time it will have grown underneath the foot and can contain quite a lot of crud.

3. Next, **carefully cut off any** ▶ **hoof material that has overgrown the fleshy part of the foot.** If trimming has been neglected, the hoof might be curved under the foot.

Helpful Hint

Dry hooves can be very hard. They're easier to trim after the goat has been walking in wet grass, and shears, which offer leverage, cut hard hooves more easily than a knife.

4. Trim off excess toe. The point of ▶ the hoof often wears down less than the sides and will require extra trimming. Heels seldom need trimming, but check them and cut very carefully, if necessary.

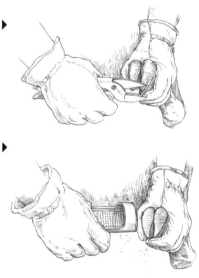

5. The final step, which is ▶ optional, is smoothing the hoof with a woodworker's Surform. The portion near the toe invariably needs more of this finishing work than the heel. You can cut quite safely until the white portions within the hoof walls look pinkish. Plane off only the horny hoof material, not the fleshy part of the foot!

6. Let the goat stand on the hoof, and see how it looks. A goat with good hooves stands squarely. A kid a few weeks old has the ideal hoof you're aiming for.

7. Do the other hind hoof the same way.

8. To do the front legs, squat down beside a front leg, bring the foot up over your knee, and trim as you did each hind hoof.

In extremely bad cases where the goat looks like it's wearing pointed elf's shoes, it may take several trimmings to get them back in shape. In such cases it's better not to cut too much at once. If the hooves are really bad, the goat isn't likely to stand around patiently while you finish. Take off as much as you can and come back a day or two later.

Don't forget: Bucks have hooves, too! These poor fellows are more likely to be neglected than the girls are, but of course they shouldn't be.

Disbudding

The other major grooming duty for goat owners is disbudding. Much has been written about the advantages and disadvantages of horns, and even more has been said around goat barns. Here's a summary of the arguments on both sides:

FOR	AGAINST
◆ Horns are protection against dogs and other predators. ◆ Horns serve as a "radiator" to help cool the animal. ◆ Horns are beautiful and natural and a goat doesn't look like a goat without them. ◆ Disbudding is ghastly.	◆ Horns are dangerous to other goats and to people, especially children. ◆ It's impossible to build a decent manger that will accommodate a nice set of horns. ◆ Horns are a disqualification on show animals (except for Pygmies). ◆ Horns aren't really much protection against dogs — look to better facilities instead, or perhaps a guard dog or llama.

One thing everyone agrees on: disbudding is much better, and easier, than dehorning.

Disbudding involves destroying the horn bud on a very young animal, before the horns really start to grow. Dehorning, on the other hand, is the surgical removal of grown or growing horns. Dehorning can be quite painful and even dangerous to the goat, and so upsetting to the surgeon that even many trained veterinarians won't do it, or at least not more than once. They certainly don't solicit the business. It's not a job for the casual or beginning goat raiser.

Disbudding is relatively quick, easy, and painless, although it might not appear so to the neophyte. The best time to disbud a goat is when it's a few days old.

The Disbudding Iron

The recommended tool is the electric disbudding iron. Kid-sized disbudding irons are available from goat supply houses, or you can make one from a large soldering iron that has a point about the size of a nickel. Grind the point until it's flat.

electric disbudding iron   modified soldering iron

An electric disbudding iron is used to cauterize the horn buds of young kids, so the horns never grow. This is much easier — on you and the goat — than removing horns after they have developed. A soldering iron can be modified by grinding down the point. This is heated in a very hot fire, like the old-time branding irons.

How to Disbud with an Iron

1. Get the iron hot enough to "brand" wood with little pressure.

2. Hold the kid on your lap. (If you'll be doing a lot of disbudding, eventually you'll want to construct a kid holding box, which is also handy for tattooing and other tasks. See page 119.)

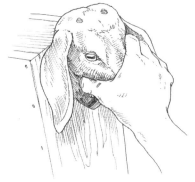

3. If the horn has not yet erupted or you're not too sure of yourself, trim the hair around the horn button with a small scissors. Some people do this routinely.

If you have more than a few kids to disbud, or don't have a helper who can hold them, a disbudding box to restrain the goat can be helpful.

4. Holding the kid firmly by the muzzle, press the hot iron into the button, and hold it there to a count of 15. There will be acrid smoke from burning hair, violent struggling (which isn't too violent with a kid weighing 10 to 12 pounds), and maybe some screaming. But when the 15 seconds are up everything will be back to normal, except maybe your heartbeat.

5. Console the kid and compose yourself while the iron heats up again, then do the other horn button.

6. After it's all over, offer the kid a bottle of warm milk and she'll forget all about it. And remind yourself that the next one will be easier.

I knew one hardy, homesteader-type lady who didn't have electricity and who heated a metal rod in her woodburning stove for disbudding.

Caustic

Another method, less hair-raising but also less successful and potentially more dangerous, is to burn the horn buds with a caustic dehorning paste. Several types and brands are available from farm-supply stores and mail-order houses.

How to Disbud with Caustic

1. Clip the hair around the horn ▶ buttons, as shown by the dotted lines.

2. Cut disks of adhesive tape to cover the buttons.

3. Apply petroleum jelly around the buttons to protect the skin from the caustic.

4. Remove the tape, and apply the caustic to the horn buttons.

5. Isolate the kid for half an hour so other kids won't lick the caustic and the treated kid doesn't get the caustic on other kids or other parts of its body by rubbing.

Caustic can cause blindness if it gets in the eyes, and it will be quite painful on other parts of the body. One lady I know holds the kid on her lap while watching television for the half-hour.

The directions on caustic are written for calves. Ignore details about when to do the job.

Caustic might seem easier, or less traumatic, than the hot iron, for the person performing the operation, and the hot iron might seem to be cruel and unusual punishment for the kid. In reality, the iron is more humane. Allowing the horns to grow might well be the cruelest alternative, if those horns some day tear a gash in another animal or put out an eye.

Breeding for Hornlessness

Some goats are naturally hornless, or polled. So many people ask, Why not breed goats for hornlessness?

One of the main reasons has been the genetic link between hornlessness and hermaphroditism. Many goats born of hornless-to-hornless matings are hermaphrodites, or of both sexes, which from a practical standpoint means they are sexless. Horned or disbudded goats can produce polled offspring, but because disbudding generally takes place before the horns erupt, these naturally hornless kids are usually disbudded anyway.

Scurs

Bucks have more stubborn horn buds than do does, and there is also a difference in breeds. If *scurs*, or thin, misshapen horns, start to develop after disbudding, merely heat up the iron and do the job over again. In some ways, scurs are more dangerous and troublesome than horns are. They can curve around grotesquely and grow into an animal's head or eye, and thin ones will be broken off repeatedly, resulting in pain and loss of blood.

Dehorning

In some cases it might be necessary to dehorn a goat. For example, if you have a herd of hornless goats and bring in a new animal with horns, she's sure to cause problems.

Grown horns can be sawed off, usually with a special wire blade. They must be removed close to the skull, actually taking a thin slice of the skull with it, or the horns will grow back. There will be a great deal of blood, and obviously a mature animal is more difficult to control. An anesthetic will be required. This is a job for a veterinarian and, as mentioned, most of them don't want to tackle it.

Again there is an alternative, which, while it might sound easier and more humane, actually isn't.

If very strong rubber bands are placed tightly around the base of the horn, the horn will atrophy and fall off. Some people file a notch in the horn very close to the skull to keep the band down where it belongs, and others claim putting tape over the band holds it on. In any case, check the rubber bands regularly to make sure they haven't broken or moved.

The problems arise when the horn structure begins to weaken. A goat may butt another, or merely get the horn caught in a manger or other obstacle, and break it off. If it's not really ready to fall off, there will be considerable pain and a great deal of bleeding. Stop the bleeding immediately.

With proper management, perhaps isolation of the animal so treated and the removal of all obstructions and very frequent inspection, the rubber-band method seems preferable over sawing, in most cases. But disbudding the young kid is far better, for all concerned.

Tip

Always keep blood-stopping powder in your barn medicine cabinet. If you don't have any, use a handful of cobwebs in an emergency.

Tattooing

Tattoos are permanent identification numbers that can help your record keeping, they can identify a goat long after you sell it, and in some cases they have helped retrieve lost or stolen animals. If you raise registered animals, you'll have to tattoo them. If your goats aren't registered, tattooing is still a good idea.

Tattoos can record a great deal of information, including the animal's age, which is indicated by a letter of the alphabet signifying the year of birth, as designated by the breed associations. For example, the letter for 1999 was M, 2000 is N, and 2001 is P, and so on. The letters G, I, O, Q, and U aren't used to avoid confusion. (By the way, if you have trouble reading a tattoo, try holding a flashlight behind the ear in a darkened building.)

tattoo set

Tattoo sets are available from farm-supply stores, mail-order houses, and small-animal equipment dealers.

Get a ¼- or ⁵⁄₁₆-inch die. Use green ink, as it shows up even on dark-colored animals.

Tattoo goats soon after birth. You need a helper or a kid-holding box (see below) to accomplish this. If it's necessary to tattoo an older animal, fasten it in a stanchion or milking stand. Where do you place the tattoo? Most goats are tattooed in the ears, though LaManchas, because they have no ears, are tattooed in the tail web.

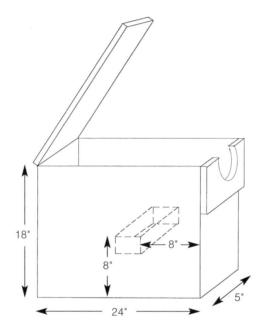

This kid-holding box, made of ¾-inch plywood, keeps the kid still for disbudding and tattooing.

You Put It Where?

"Tattoo in the ears?" people asked Judy Kapture, *Countryside* magazine's former goat editor. "Doesn't that hurt?" Judy just waggled her dangling, pierced earrings at them and smiled!

How to Tattoo

1. **Clean the area to be tattooed** with a piece of cotton dipped in alcohol or carbon tetrachloride. Stay away from warts, freckles, and veins.

2. **Smear a generous quantity of tattoo ink over the area.** Paste ink can be applied from the tube; liquid ink requires a small brush, such as a toothbrush.

3. **Place the tattoo tongs in position, and puncture the skin with a firm, quick squeeze.** (Be sure to test the numbers on a piece of paper first; they're backward, like printer's type.) On very thin-eared kids the needles might go all the way through the ear. Gently release the ear, and pull the skin free. Some tattoo tools have an ear-release feature that eliminates this problem.

4. **Put some more ink on the tattoo area, and rub it in thoroughly with the small brush.** It takes about a month for the tattoo to heal thoroughly.

The tattoo contains the letter-designated year of birth and additional information, such as the herd identification number assigned by a registry.

MATERIALS & EQUIPMENT
Cotton balls
Rubbing alcohol or carbon tetrachloride
Tattoo ink
Small brush or toothbrush
Tattoo set

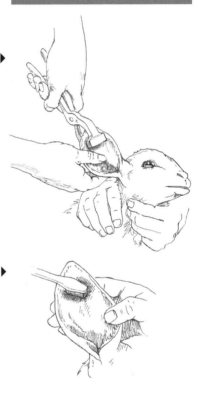

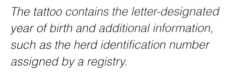

Hair Care

People who work around food wear hairnets. Since that's not practical for goats, trimming and brushing are important for high-quality milk.

Very shaggy goats should have their hair clipped, especially around the flanks and udder. It's a good idea to confine trimming to the udder region in the winter, to keep your goat warmer, but even that "just a little off the top" will keep your milk cleaner. The entire animal can be clipped in the spring to keep it cleaner and cooler and to discourage parasites. Just like a haircut, a nice clipping will greatly enhance the appearance of a show goat.

Electric clippers are nice, but hand clippers (such as those for dogs) work well. If you only have one or two goats and don't want to invest in clippers, and can't borrow any, sharp scissors will do, but be very careful.

HEALTH

*The desire to take medicine is perhaps the greatest feature
which distinguishes man from animals.*

— Sir William Osler, Canadian physician, 1849–1919

Goats are among the healthiest and hardiest of domestic animals. If you pay close attention to proper feeding and other management details, you're likely to have very few health problems with your goats. This chapter gives you a brief A-to-Z introduction to some of the health problems you might hear about or encounter.

Health: The Natural State

It seems to me that some people are obsessed by disease and sickness. They waste money, worry unnecessarily, and don't enjoy life with goats as they should. My own views are different, and I'll explain my attitude so you'll be aware of my bias. Then, if you don't agree, perhaps you can find a more technical manual, written by a veterinarian, or at least by someone who shares your interest in sickness. (*Goat Medicine*, by Mary C. Smith and David M. Sherman, published by Lea & Febiger, 1994, is highly recommended.)

In my view, sickness is only an absence of health, and health is the natural state. If your animals get sick, it's because of wrong conditions of feed, environment, or in some cases breeding. Treating the symptoms will help in the short run, sometimes, but unless the underlying causes are corrected, any time and money spent on medication is wasted.

What's worse, many illnesses have purposes, and by "curing" them we sometimes compound the problem. Scours or diarrhea is one example. It's fairly common in kids, and can result from feeding too much milk, cold milk (when the kid isn't used to it), or dirty utensils. You don't want to stop the diarrhea cold because that's nature's way of getting rid of the toxins. So you let it take its course while removing the *cause*, the excess milk, the cold milk, or the unclean utensils. Similarly, a completely worm-free goat is a near impossibility and not a desirable goal under any circumstances. The number and amount of vermifuges (dewormers) required would do more harm than good, and some internal parasites are symbiotic — the goat needs them to live.

We have been led to believe that all microorganisms are bad per se. Nothing could be further from the truth. Even most pathogenic organisms will have little or no effect on a healthy body; only when the host is weakened because of some other factor, such as poor nutrition, does the pathogen get out of hand. Some bacteria are apparently harmless, and some are actually necessary.

Your job, then, is to maintain the natural state of your goat's health by providing her with the proper feed and environment.

Finding and Using a Veterinarian

This isn't meant to imply that goats never get sick, that if they do it's because we did something wrong, or that there's nothing we can do for them. While you might go for years without seeing any health problems, if you have a large number of goats (or live with a few of them long enough), you're almost certain to encounter some ailments. However, you don't need a medical degree to raise goats. If an animal gets sick, all you need is the phone number of a veterinarian.

One of the most common complaints I hear from goat people is, "My vet doesn't know anything about goats." It's true that a few vets just don't care about goats. Others simply lack experience with them due to their low population and generally good health. In any event, anyone who has graduated from veterinary school knows more about animal diseases than the rest of us do. I believe in making use of their knowledge. And if you have a good working relationship with a capable veterinarian, he or she will be glad to share much of that knowledge with you.

Do try to find a large-animal veterinarian, one who treats sheep and cattle, not cats and dogs or pet birds. And it's a good idea to start a professional relationship by engaging a veterinarian for help or advice concerning routine health maintenance *before* you desperately need help at 2 o'clock on a stormy morning.

Basic Physiological Data

PARAMETER	NORMAL
Temperature	101.5–104.0°F (rectal). Varies with air temperature, exercise, excitement, and amount of hair. To determine abnormal temperature, compare with several others under same conditions.
Pulse	70–80 per minute
Respiration	12–20 per minute
Puberty	Occurs at 4 to 12 months
Estrus (heat) cycle	18–23 days
Length of heat period	18–24 hours (average); range 12–36 hours
Gestation	148–153 days; average 150 days
Birth weight	8 pounds

An A-to-Z Guide to Common Health Problems

Here's a brief overview of some of the conditions you're most likely to encounter, with tips on treatment and prevention.

Abortion

Research has indicated that Bang's disease (*Brucella abortus*) is extremely rare among goats in the United States, but Bang's disease tests are commonly required for showing, shipping, and for selling milk. Goats do abort, however. (See Bang's Disease, page 127.)

If abortions occur early in pregnancy, the cause is apt to be liver flukes or coccidia. Liver flukes are a problem primarily in isolated areas such as the Northwest where wet conditions favor them. Coccidia can be transferred by chickens and rabbits, both of which should be kept away from goat feed and mangers. (See Coccidiosis, page 130.)

Abortion is more common in late pregnancy. The cause can be mechanical, such as the pregnant doe being butted by another or running into an obstruction such as a manger or a narrow doorway, or it can be related to moldy feeds.

Certain types of medication can cause abortion, including worm medicines and hormones, such as those contained in some antibiotics. Medicate pregnant animals with caution.

Abscess

An abscess is a lump or boil, often in the neck or shoulder region, that grows until it bursts and exudes a thick pus. There are several types, with different causative organisms, and very different degrees of seriousness. Most are related to wounds, including punctures by thorny vegetation, bites, and even hypodermic needles. An abscess can occur when a goat bites her cheek. But the form getting the most attention by far nowadays is *caseous lymphadenitis*, or CL. It's caused by *Corynebacterium pseudotuberculosis*, formerly known as *Corynebacterium ovis*.

The condition can be transmitted from one animal to another. Therefore, abscesses are common in some herds and nonexistent in others.

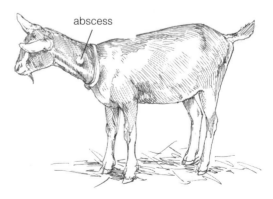

abscess

This doeling has an abscess on her neck. Although this condition is quite common in some herds, it does require special attention.

First isolated in sheep in Australia in 1894 (hence the name C. *ovis*), CL was seldom mentioned in the United States until the 1970s, and it didn't reach England until the 1980s, with the importation of Boer goats. Today, many consider CL to be the major disease problem of dairy goats in the United States. The late Dr. Samuel Guss, a well-known and highly respected authority on goat health, called a goat with a discharging abscess "a hazard to other goats and to humans."

Treatment

Any animal with an abscess should be isolated. The milk from such an animal should be pasteurized, according to Dr. Guss, and if the abscess is on the udder the milk should be discarded. Don't feed it to kids; dump it.

The lump can become the size of a walnut, a tennis ball, or even larger. It will burst by itself, thereby spreading the infection. To avoid this, the abscess should be properly lanced and cleaned before it ruptures.

Clip the hair around the abscess, and disinfect the surface with an antiseptic such as tincture of iodine. Make a vertical incision, as low as possible to promote drainage, with a sharp, sterilized knife. Wearing disposable latex gloves, squeeze out the pus and burn the material and all cloths that come in contact with it. Flush the wound with dilute disinfectant. Isolate the animal until the lesion is healed and covered by healthy skin, typically 20 to 30 days.

If the abscess is in the throat behind the jaw or under the ear, a veterinarian should perform the procedure, since these areas contain major blood vessels and nerves. But there are other reasons to engage expert help.

If the abscess is caused by *lymphadenitis* it will have a greenish cheese-like pus. A yellowish pus of mayonnaise consistency suggests C. *pyogenes*, while *streptococci* often produces a watery discharge and *staphylococci* causes a creamy exudate. However, these are only clues. The only way to know for certain is with a lab culture.

Prevention

There has been some work, and obviously, a great deal of interest, in vaccinations, particularly in Australia and Canada. However, success has been variable and no preventive is currently on the market in the United States. Control consists mainly in treating and isolating (or culling) infected animals. Remove kids from affected does at birth and raise them on colostrum and milk from clean does.

Herds that are free of abscesses generally stay that way until a new animal is brought in or the goats come into contact with others in some other way. But once established in a herd, CL is difficult to eradicate.

Bang's Disease (Brucellosis)

Brucellosis is a contagious disease primarily affecting cattle, swine, sheep, goats, and dogs, and characterized chiefly by abortion. The first *Brucella* infection to be recognized was caprine brucellosis, or Malta fever, in 1887. While it's prevalent in many countries where goats are common, brucellosis is extremely rare in the United States. The last outbreak was in a herd of meat goats in Texas in 1969. But it's still considered a dreaded disease, and goats are routinely tested for it.

Abortion occurs in about the fourth month of pregnancy. Diagnosis requires a bacteriologic examination of the milk or the aborted fetus, or a serum agglutination test. There is no cure: the disease is treated by the slaughter of reacting animals.

Many people are concerned about brucellosis, but several years ago *Countryside* magazine investigated the thirteen cases reported in U.S. Department of Agriculture (USDA) annual statistics. These cases occurred in three herds, in Arizona, Indiana, and Ohio. A check with officials in Arizona and Indiana showed that their cases were in fact clerical errors. The one goat in Ohio was classed as positive on the test, suspicious on the retest, and after being slaughtered and subjected to a tissue test, negative.

Goats that have come up suspect after a Bang's test have invariably been pregnant or recently freshened. Subsequent tests are negative.

Nevertheless, to be absolutely certain about the safety of the milk from your goats, you can have them tested for Bang's, and you can pasteurize the milk.

Bloat

Bloat is an excessive accumulation of gas in the rumen and reticulum resulting in distention. Bloat is caused by gas trapped in numerous tiny bubbles, making it impossible to burp. If you've just turned the goat out on a lush spring pasture or if she found out how to unlock the door to the feed room, anticipate bloat.

Treatment

A cup of oil — corn, peanut, or mineral — will usually relieve the condition. A handful of bicarbonate of soda will help. In extreme cases it may be necessary to relieve the gas by making an incision at the peak of the distended flank, midway between the last rib and the point of the hip and holding the wound open with a tube or straw.

Prevention

As always, the best cure is prevention. Feed dry hay before letting animals fill up on high-moisture grasses and clovers. Don't feed great quantities of succulents such as green cornstalks if the animals aren't used to them.

Alternate Bloat Remedy

Give a mineral-oil enema, followed by lots of water.

Caprine Arthritis Encephalitis (CAE)

In the early 1970s, CAE was little more than a goat-world rumor. While a few breeders became almost hysterical, most thought it was much ado about nothing. By the early 1990s, some researchers claimed that 80 percent of the goats in the United States were infected with CAE virus, or CAEV. That scared almost everybody.

Subsequent studies on larger numbers of goats in more diverse herds brought that down to 31 percent. Dr. Joan Bowen, a well-known goat veterinarian, says that in her experience only 10 percent of these will develop clinical disease. This would mean that about three goats out of a hundred will develop clinical symptoms of CAE. Another problem is that most currently available tests don't indicate the presence of CAE virus, but only the *antibodies* to the virus. The most common CAE tests — agar gel immunodiffusion (AGID) and enzyme-linked immunosorbent assay (ELISA) — are said to have only an 80 percent accuracy, and the most accurate tests are expensive.

Symptoms

In one common form of the disease, the first signs are usually minor swelling in the front knees. The swollen knees become progressively worse, and the animal just seems to "waste away." The lungs may become congested, and eventually all the body's systems give up. Other symptoms involve chronic progressive pneumonia, and weight loss associated with chronic disease. Less common is an ascending paralysis in kids that otherwise appear healthy.

Treatment and Prevention

Caprine arthritis encephalitis is an incurable contagious disease, and some goats that have the virus do not show symptoms but are still carriers. There is no vaccine. The only recourse at this time is a prevention program.

You can have your goats tested for CAE. If they test positive, there is nothing you can do for them, but because the infection is spread in the neonatal period, you can build and protect a "clean" herd by following this regimen:

1. Be there when the kids are born. Deliver them onto clean bedding, preferably newspaper rather than straw. Don't break the amniotic sac before the kid is delivered: the fluid in the sac is not infected, and the sac prevents the kid from swallowing or inhaling infected cells. Do not let the doe lick the kid.

2. Put each newborn kid into clean, separate boxes so they can't lick one another. As soon as you can, wash each one in warm, running water, to eliminate the possibility of any infected fluid on the body being ingested. Keep them separate until they're clean and dry.

3. Within half an hour or so, feed the kids colostrum — from a goat you know is free of the virus, or from a cow, or in case of emergency, a commercial or home-brewed colostrum replacer. (See page 171.) Note that cow colostrum has recently come under fire because of the possibility of introducing other diseases, such as Johne's (see description on page 132).

4. Feed only pasteurized milk, cow milk, or sheep or goat milk replacer. (Milk replacer made for calves is not high enough in fat for sheep or goats.)

5. Keep the kids separated from other goats and practice strict hygiene, including not only sanitized feeding utensils but also such precautions as caring for the kids before handling or walking among the older animals, and washing your hands, changing boots, and so on before going among the kids.

Should You Worry?

How concerned should you be about CAE? That depends. If a homestead dairy raises its own herd replacements or uses all kids for meat, a CAE prevention program isn't likely to be profitable, or even necessary. The more sophisticated tests can cost as much as $50 per sample, which is not cost effective for most backyard dairy goat owners. On the other hand, breeders who sell goats (particularly across state lines) or who attend shows that require health certificates should use a CAE prevention program as outlined above.

Caseous Lymphadentis

See Abscess, page 125.

Coccidiosis

This disease is caused by microscopic protozoans (*coccidia*) found in the cells of the intestinal lining, and is therefore a parasitic disease. (See also Parasites, Internal, page 136.) It usually occurs in kids 1 to 4 months old, and usually in crowded and unsanitary pens. The most common symptom is bloody diarrhea, although kids with coccidiosis are usually weak and unthrifty. Your vet might recommend a sulfonamide in the feed mixture to treat this condition.

Cuts

Cuts, punctures, gashes, and other wounds can almost always be avoided by good management. They can be caused by such hazards as barbed wire, horned goats, junk or sloppy housekeeping and other conditions under the control of the goat caretaker. Still, accidents can happen.

Clean such wounds with hydrogen peroxide and treat with disinfectant such as iodine. Use your own judgment to decide if stitching is required, or get the animal to a veterinarian. Verify that tetanus vaccination is current.

Diarrhea

See Scours, page 140.

Enterotoxemia

Enterotoxemia is also called pulpy kidney disease and overeating disease. An autopsy soon after death will often show soft spots on the kidney.

The usual symptom of enterotoxemia is a dead kid. There is always misery, and almost always a peculiarly evil-smelling diarrhea. With some strains, there may be bloat or staggering.

It is caused by a bacterium that is always present, but that, when deprived of oxygen in the digestive system, produces poisons. There are six types of *Clostridium perfringens* bacteria that cause enterotoxemia. Types B, C, and D cause the most trouble, with type D most often affecting sheep and goats. The proper conditions can be induced by overfeeding. Goats build up resistance to the poisons produced in small, regular amounts, but they can't handle sudden surges of them.

Treatment

Antitoxin can be administered if you get there fast enough, but death is usually swift. Where enterotoxemia is a problem, vaccines are available from your veterinarian. Annual booster shots are required, and the kids will get antibodies from their dams.

Prevention

The best prevention is proper feeding on a suitably bulky, fibrous diet.

Goat Pox

The symptom is pimples that turn to watery blisters, then to sticky and encrusted scabs on the udder or other hairless areas such as the lips. It varies in severity.

Pox can be controlled by proper management, especially involving sanitation. Infected milkers should be isolated and milked last to avoid spreading the malady to others. Time and gentle milking are the best cures. Traditional treatment is methyl violet to dry up the blisters, but this is very drying and can make the udder painful.

Prevention

Very similar conditions can be caused by irritation. I have seen cases caused by dirty, urine-soaked bedding and by the use of udder-washing solutions that were too strong. In these cases the cure is wrought by removing the cause. An antibiotic salve will keep the skin supple and prevent secondary infections.

Johne's Disease (Paratuberculosis)

Like CAE, this is one of the "wasting" diseases. It primarily affects the digestive tract, probably with fecal-oral transmission. Frequently the only symptom is extreme weight loss or gradual loss of condition. Scouring is a typical symptom in cattle but uncommon in goats. It cannot be diagnosed accurately in goats except by autopsy. The intradermal johnin test used on cows has come up negative on goats that actually had the disease as determined by autopsy.

The disease apparently infects young animals, either by interuterine transmission, congenitally at birth, or by mouth. If the disease follows the pattern it does in cows, adult animals can be sources of infection even if they do not show clinical signs of disease. The kid infected at birth typically won't start getting sickly for 1½ to 2 years.

There is no reliable test for goats and no cure. A vaccination developed in Australia has not been approved by the U.S. Food and Drug Administration and so is not available in the United States.

Prevention

The best prevention consists of starting with a clean herd and keeping it that way. Beyond that, infected individuals should be identified and removed from the herd, and new infections in susceptible kids should be reduced by improved sanitation and modified kid-raising methods, including isolation from adults. Do not let the doe lick the newborn, and don't allow the kid to nurse.

Ketosis

Ketosis includes pregnancy disease, acetonemia, twin-lambing disease, and others. Symptoms include a lack of appetite and listlessness. Ketosis occurs in the last month of pregnancy, or within a month after kidding. Its primary cause is poor nutrition in late pregnancy, but it's most likely to affect fat does, especially those that get little exercise. A dairy goat should never be fat; nutrition is particularly important when the unborn kids are growing rapidly and making huge demands on the doe.

Treatment

Treatment consists of administering 6 to 8 ounces of propylene glycol. This may be given orally twice a day, but not for more than 2 days. In an emergency situation, try a tablespoonful of bicarbonate of soda in 4 ounces of water, followed immediately by 1 cup of honey or molasses. Once the advanced stages — characterized by inappetence, then complete anorexia — have developed, no treatment is effective.

Lice

Suspect lice if your goat is abnormally fidgety and has a dull scruffy coat. Fresh air and rain are good preventives. For an old-time cure, apply two parts lard to one part kerosene.

Lice are almost universal, but mild infestations cause little harm to well-nourished animals. A badly infested goat will rub against posts and other objects, have dry skin and dandruff, and can lose a great deal of hair. Lice can be controlled by dusting, spraying, or in large herds, dipping. Ask your veterinarian for a louse powder approved for use on dairy animals. All members of a herd must be treated at the same time to control lice. With plenty of sunlight, fresh air, and rain, lice seldom become a serious burden.

Mange

Mange is indicated by flaky, scurfy, "dandruff" on the skin. It's accompanied by irritation. Hairlessness develops and the skin becomes thick, hard, and corrugated. The condition is caused by a very tiny mite. There are several types of mange. Demodectic is probably most common and can be stubborn. Mange can be treated with a variety of medications, including amitrol and lindane, available from veterinarians. Follow the directions on the label.

Scurfy skin can also be the result of malnutrition or internal parasite infestation.

Mastitis

Mastitis is inflammation of the mammary gland, usually caused by an infection. Symptoms are a hot, hard, tender udder; milk may be stringy or bloody. (Routine use of a strip cup before milking will alert you to abnormal milk; see chapter 13 for details.) Mastitis may be subclinical, acute, or chronic. It's usually a relatively minor problem, but some forms, such as gangrenous, can be deadly.

Not all udder problems indicate mastitis. Hard udders (typically just after kidding) that test negative for mastitis are referred to as congested, and usually disappear. Congested udders are best cleared up by letting the kids nurse, and massaging the teats and udder for 3 to 4 days after parturition. In mastitis, the alveola or milk ducts are actually destroyed. Since it's necessary to identify the bacteria involved, the services of a vet are required.

Mastitis can be caused by injury to the udder, poor milking practices, or by transference by the milker from one animal to another. Hand milkers should wash their hands between each animal, and udders and teats should be washed and dried before milking. Teat dips have proved of great value in controlling the disease among cattle, although the solutions must be diluted for goats.

Somatic cell counts, used to detect mastitis in cows, are not considered reliable with goats. (See page 254.)

Home tests for mastitis are available from veterinary supply houses. The best-known is the California mastitis test, or CMT. However, Smith and Sherman in *Goat Medicine* say, "The CMT is more useful in ruling out than for diagnosing mastitis in goats. In a well-managed herd, the predictive value of a positive test is unacceptably low."

Use a Teat Dip after Each Milking

Goat veterinarian Dr. Joan Bowen says that consistent use of an effective teat dip after every milking could prevent half of all cases of mastitis. She explains that during milking, the small sphincter muscle around the teat orifice relaxes, allowing the orifice to open and the milk to drain out. It takes about 30 minutes for this orifice to close again. This is when bacteria can enter the teat canal, causing mastitis. A teat dip kills the bacteria before the orifice closes.

Prevention

As you might expect, an ounce of prevention is worth a pound of cure. Dry, clean bedding, especially at freshening, is important. Wash and dry udders with individual towels for each animal. Use a spray bottle for the sanitizer solution, rather than a communal bucket. Milk with clean, dry hands. Milk gently, in peaceful surroundings. Avoid vigorous stripping.

Mastitis and Length of Milking Intervals

You might be interested in knowing that the widespread belief that mastitis can result from not milking at 12-hour intervals has not been proven. One study has shown that goats milked at intervals of 16 and 8 hours produced as much as those milked at 12-hour intervals, and with no increase in mastitis. Milking should be done at regular times, but it doesn't have to be a 12-hour schedule.

Milk Fever (Parturient Paresis)

Symptoms of milk fever include anxiety, uncontrolled movements, staggering, collapse, and death. Usually this occurs within 48 hours of kidding. It's caused by a drastic drop in blood calcium, which is related to the calcium level of feed consumed during the dry period, and even to incorrect feeding of young animals. It can be brought on by sudden changes in feed or short periods of fasting. Curing milk fever requires quick action and a veterinarian, who will administer calcium borogluconate intravenously, which is typically quite effective.

Parasites, External

External parasites are generally much less of a problem with goats than are internal parasites.

Screwworms can be a problem, particularly in the South and Southwest. They survive on living flesh, and normally depend on wounds — including those from dehorning and castration — in order to enter an animal. Castrating with the Burdizzo (see chapter 9) and timing dehorning so the wounds will be healed before fly season will help prevent screwworm infestations.

Dog and cat fleas seldom bother goats, and then only in tropical regions. Goats can be affected by mites, which produce the diseases mange and scabies. Sarcoptic mites, responsible for sarcoptic mange, can affect all species of animals; demodectic mange and psoroptic ear mange are specific to goats. Demodetic mange is generally associated with crowded confinement housing, as the mites survive only a few hours away from the goats.

See also Lice, Mange, page 133.

Parasites, Internal

"Worms" of various kinds are perhaps the most widespread and serious threat to goats' well-being, but only when they're present in large numbers. Many goat raisers worm goats regularly with a favorite anthelmintic, or wormer, but, alas, it isn't quite this simple.

The list of internal parasites that infest goats is quite long. It includes bladder worms, brown stomach worms, coccidia, four species of *Cooperias*, hookworms, liver flukes, lungworms, nodular worms, stomach worms, tapeworms, whipworms, and others. Some are quite common in certain areas and rare elsewhere.

Not all of these are affected to the same degree by a specific anthelmintic. This means a fecal test is required so your veterinarian (or you, if you have a microscope and an interest in such matters) can determine which parasites are present, and therefore which veterinary product to use.

Deworming Agents

Some of the more common vermifuges include levamisol (Tramisol), thiabendazole (TBZ; Omnizole), cambendazole (Camvet), fenbendazole (Panacur), mebendazole (Telmin), and oxfendazole (Benzelmin). As if matters weren't complicated enough already, some cross-resistance to some of these drugs in the same chemical family has been reported.

The best method of dealing with parasite problems or potential problems consists of two simple steps: (1) regular and close inspection of your goats, and (2) periodic fecal exams.

Inspecting Your Goats

Pay attention to your goats' mucous membranes, the gums, and particularly the eyes. The "whites" of the eyes should not be white, but pinkish or red. If they're white, or if the gums are pale pink or gray, this indicates that the goat is anemic. The likely cause is worms.

This isn't foolproof either, though. *Muellerius*, for example, doesn't cause anemia; it destroys lung tissue. (Also, it is not affected by the most common goat wormers. If you simply worm on a regular basis, you might enjoy a false sense of security.) In addition, many parasites can build up resistance to a given anthelmintic. It's usually necessary to periodically rotate the products used. Consult a veterinarian for specific recommendations.

Important Note

Few medical products, such as wormers, are officially approved for goats, because of the expense of testing and the small market potential for manufacturers. In addition, as with antibiotics in humans, drug resistance is showing up with worrisome regularity in goats. That's why it's best to work with a veterinarian who has the latest information on medications, including *withdrawal times* (that is, the period between administering the product and being able to safely consume the milk or meat). And always be sure to read product labels!

Methods of Administration. Anthelmintics come in various forms: boluses (large pills), pastes, gels, powders, crumbles, and liquids. Boluses are popular, but many goats refuse to take them, and because they can choke a goat, some people refuse to use them. They can be administered with a balling gun, or try hiding a bolus inside a gob of peanut butter.

Drenching, or administering a medication from a bottle, can also be risky, but it's almost a required goatkeeping skill. It's important to: (1) give a little at a time and allow the goat to swallow in-between; (2) give in the left-hand corner of the mouth; and (3) never raise the head — keep the muzzle level. The risk is getting the medication in the lungs.

When administering paste wormers, be sure the goat doesn't have anything in its mouth, including a cud. Put the paste in the back corner of the mouth on the left side. If the goat wants to shake her head and fling the paste out of her mouth, hold her muzzle gently and massage her throat until you're sure the wormer has been swallowed.

Fecal Exams

The best advice that can be given to beginners, or to anyone who doesn't want to become an expert on worms, is to have laboratory fecal exams performed twice a year, and to follow the advice of a veterinarian. A schedule can be set up based on the life cycles of the specific parasites present and the anthelmintics chosen.

The Better Approach

You can deworm your goats on a schedule because books, articles, and other breeders say that's what you should do. Or, you can have your veterinarian check your goats to determine whether they have worms; and if so, what kind; and then take appropriate action. The latter approach is cheaper in the long run and easier on, and better for, both your goats and you.

Incidentally, most people call it *worming,* but *feeding* and *watering* refer to giving goats feed and water, and we sure don't want to give them worms! *Deworming* is the better term.

Pneumonia

Pneumonia is a broad term referring to a number of lower respiratory tract diseases. The etiology (origin) is often unknown without a report from a microbiologist or pathologist. Until that report is in hand, even a veterinarian often has to make an educated guess at a specific diagnosis and, therefore, about antibiotic therapy. This is usually based on experience, but if one antibiotic doesn't work, another may be used. In addition, multiple factors are often present, including worms. What all this means is that for the goat owner, the main thing to know about pneumonia is how to prevent it.

Prevention

Ventilation tops the list. Goats needn't be kept in warm housing, but it *must* be well ventilated and draft free. In a closed barn, an engineered ventilation system can be considered essential. In cold climates, newborn kids can be removed from the main barn to more suitable housing, or if left with their dams, dressed in wool coats or body socks (often made from discarded sweatshirts or similar clothing). Plenty of colostrum and dry bedding are needed.

Crowding must be avoided. Wet bedding must be avoided. High humidity and condensation must be avoided. Insulation can reduce ceiling condensation but will increase the need for ventilation. *Never* line a barn with plastic sheeting, which will increase the humidity. Repair leaking waterers that add to humidity.

Dry, clean bedding is essential, but minimize dust. Isolate new or stressed animals, including those that have traveled to shows.

Poisoning

Symptoms are vomiting, frothing, and staggering or convulsions. Because of the nature of a goat's eating habits, poisoning from plants is rare. The goat takes a bite of this and a taste of that and will seldom eat enough of one poisonous plant to cause much damage. Toxic plants are more common in the western states than elsewhere. To learn what plants in your particular locale are poisonous, check with your local county extension agent. Some to watch out for are locoweed, milkweed, wilted wild cherry leaves, and mountain laurel. (See chapter 6.)

Lead poisoning used to be a possibility when goats chewed on painted wood or were fed weeds gathered along roadsides contaminated with auto exhaust. Both causes are less common today with the banning of leaded paint and gas. Still, avoid taking feed of any kind from along roadsides that might have been sprayed for weed control.

Don't feed Christmas trees to goats because many of them are sprayed with toxic substances. If your neighbor sprays any crop, keep your goats away from any area that might have been contaminated by drifting spray.

Many seeds are treated and poisonous. Every so often, we hear of fertilizers or insecticides or other chemicals that look like feed additives killing off whole herds of cows when someone mistakenly grabs the wrong bag. Be careful.

Antidotes depend on the poison. Call a veterinarian.

Scours (Diarrhea)

Scours in newborn kids can indicate any of a number of problems, including failure to ingest colostrum soon after birth, lack of sanitation, inadequate nutrition of the doe during gestation, feeding excessive amounts of milk, and feeding low-quality milk replacers. The mortality rate is high, and swift action is required.

Homemade Electrolyte Solution

Since death results from dehydration and shock,
the first goal is to restore electrolyte balance.
Electrolyte formulations are available from drug companies,
but in an emergency a suitable solution can be mixed from ingredients
found in any kitchen. Here is one of many.

2 teaspoons table salt
1 teaspoon baking soda
8 tablespoons honey, white corn syrup, or crystalline dextrose
   (never cane sugar!)
1 gallon warm water
Neomycin, nitrofurazone, or chloramphenicol can be added
   to formula, or given separately

1. Add salt, baking soda, and sweetener to water.
2. Mix well.
3. Add antibiotic to formula if you wish.
4. If the kid is too weak to nurse, administer with a syringe or stomach tube. Give 1 to 2 cups per 10 pounds of body weight per day until the scours clears up. Don't feed milk during this period.

Hypotonic vs. Isotonic Solution

This electrolyte solution is called *hypotonic*. It contains electrolytes in roughly half the concentration of electrolytes in the blood. This is given only by mouth. However, a veterinarian can administer *isotonic* electrolyte solutions in which the concentration is the same as in the blood, intravenously.

Home Remedy for Scours

This is another home remedy for scours, provided by a veterinarian.

1 cup buttermilk
1 raw egg
1 teaspoon cocoa
¼ teaspoon baking soda

1. Mix ingredients in a blender, or shake well in a jar.
2. Bottle feed one-fourth of this mixture every 2 to 3 hours. One crushed bolus of neomycin can be added to this mixture.

Tetanus

Goats with tetanus, also called *lockjaw*, will have their heads held up in an anxious posture and will be generally tense. There is difficulty in swallowing liquids, and muscular spasm. Death occurs within 9 days. Tetanus requires a wound for the germ to enter, but it can be something so simple the caretaker doesn't even notice it. Disbudding, tattooing, fighting, castration (especially with elastrator bands), dog bites, and even hoof trimming can set a goat up for tetanus. Horses and mules are often associated with tetanus, but the spores are widespread in soil and animal feces, and can survive for many years.

Treatment

Treat all punctures and cuts with iodine, and pay special attention to the navels of newborn kids. Routine tetanus vaccination is recommended. Treating tetanus is a job for a veterinarian, but early identification and treatment are important.

Tuberculosis

Contrary to the fervent belief of many hobbyists, Mary C. Smith and David C. Sherman point out in their book *Goat Medicine* that goats *are* susceptible to tuberculosis. Goats may serve as a reservoir of infection for cattle or may directly infect humans.

White-Muscle Disease
(Nutritional Muscular Dystrophy)

This is caused by a lack of selenium or vitamin E. It most commonly affects healthy, fast-growing kids less than 2 months old, although problems can occur in mature animals. Diagnosis is often difficult even for a veterinary practitioner. Stiff hind legs can indicate white-muscle disease in kids, but it could be tetanus. Sudden death in young kids could be caused by white-muscle disease, or by gastrointestinal parasitism. Some cases show symptoms similar to enterotoxemia. Definitive diagnosis requires a postmortem examination of muscle lesions or examination of blood or tissues.

Soils in many parts of the United States are deficient in selenium. Providing this mineral (5 mg, orally or intramuscularly) 2 to 4 weeks before kidding will prevent deficiencies in the doe and in the kid for its first month. Kids should receive injections at 2 to 4 weeks. There is a relationship between selenium and vitamin E. The two are usually administered together.

Note that in excessive amounts, selenium is a poison. In some areas soil levels are high enough so that plants grown on them cause toxicity, resulting in paralysis, blindness, and even death.

Worms

See Parasites, Internal, page 136.

Don't Expect to Be a Goat Doctor

To repeat, if you start with healthy goats and give them proper care, chances are good that they'll have few health problems.

Not that goats don't get sick. But that's another problem: rare as they might be, there are so many *possible* diseases and ailments that it doesn't make any sense for a person with a few animals to even try to be familiar with them all, even if that's possible.

If you do have a sick goat, don't hesitate to call a veterinarian. It's easy for amateurs to misdiagnose animal illnesses. Furthermore, if you try to get a veterinary education from a couple of books, try all the home remedies you can find, and only call the practitioner when the animal has three feet in the grave, don't expect the doctor to perform miracles. Veterinarians carry drugs, not Bibles.

Available Drugs and Extra-Label Use

Most homestead goat raisers are not keen on using drugs, but those who don't share that aversion should know that only eight are currently licensed for use in goats: decoquinate (Deccox), monesin sodium (Rumensin), morantel tartrate (Rumatel) medicated premix, 10% fenbendazole (Panacur) solution, ophthaine solution, nitrofuranzone puffer, thiabendazole (Omnizole), and oral neomycin.

Certain others can be used, but only under the supervision of a veterinarian. Known as "extra-label use," conditions are spelled out in the Animal Medicinal Drug Use Clarification Act under the authority of the FDA. One rule is that a veterinarian-client-patient relationship must exist. It is strictly *illegal* for nonveterinarians to use these drugs.

THE BUCK

Goats won't produce milk without kidding and they won't kid without being bred. That requires the services of a buck, and that entails a whole 'nuther look at goat raising.

You'll need a buck, but that doesn't mean you'll have to buy one. In fact, beginners are usually advised to forget about keeping a buck. There are many reasons.

Why *Not* to Keep a Buck

One practical motivation is expense. A buck requires the same amount of housing, feed, bedding, and grooming as a doe. Therefore, if you have one doe and one buck, the cost of your milk is double what it would be without the buck. The number of does you need to justify the expense of having your own herd sire depends pretty much on your particular situation. If you live in a remote place where there is absolutely no stud service available within a "reasonable" distance (and this might be a few miles or a few hundred miles, again depending on you) then it might be necessary to have a buck even for just a few does. Expensive, but more economical than the alternatives.

A reason just as practical, in many cases, is that bucks do stink! And some bucks that were not properly raised or trained can be dangerous, especially to people not physically and mentally equipped to handle them. We'll look at both of these in more detail in a moment.

Why to Keep a Buck

But perhaps the biggest problem of all is that before you decide to keep a buck, you'll have to make a much more important and difficult decision: Why do you need a buck? The obvious answer is to breed does so they can produce kids and milk, but it's not that simple. If all you want from your goats are milk and meat, you might assume that any "doe freshener" will serve the purpose. That might be shortsighted, even if your only interest is being a self-sufficient homesteader. Here are a few reasons why you want the best buck available.

Improving the Breed

Good bucks are expensive, and serious breeders won't sell any other kind. They're worth it, of course, because the buck you use this year will affect your herd for years to come. While a doe can be expected to produce one or two kids a year, a mature buck can breed as many as a hundred does a year. If you only use one buck, his genes will be passed on to every kid you raise. The buck is truly "half the herd."

A good breeder won't sell you a "cheap" buck. If it isn't good enough to improve most goats of its breed, it is slaughtered at birth or castrated. Unfortunately, there are too many people raising goats who are *not* good breeders. They'll sell any buck, often for less than a doe. And too many beginners think this is a good deal. It might not be.

This deserves some explanation, because there are two approaches to raising livestock, and not only are most beginners somewhat puzzled by the differences, but their natural inclinations frequently attract them to the less desirable approach.

Why Bother with Improvement?

We're speaking of breed improvement. While most emphasis on breed improvement naturally comes from people who are involved in showing their animals, be they rabbits, dogs, cows, or goats, there is more than ample evidence to prove that the "commercial" or homestead or backyard producer has every bit as much to gain from striving for improvement, and perhaps more. And there is really very little — or nothing — to lose.

I have found it frustrating to deal with people who place little or no emphasis on breed improvement, or who even actively belittle the "fancy-pants" show enthusiasts as if their interests were somehow contradictory. Nothing could be further from the truth.

For proof, we need only turn to the commercial dairy (cow) farmer. Almost invariably these practical, tough-minded, cost-conscious farmers use the best purebred registered bulls they can find. They may not have the slightest intention of ever showing a cow or of raising registered cattle (although some of them are finding that registered cows are valuable for the same reasons registered bulls are practical). They use purebreds because it pays off in the milk pail. Milk production per cow has doubled since the last century. While some of this is due to feeding practices and other management details, a large share of the credit must go to genetics.

Similar progress has only recently taken place in goats. But even today, no lover of goats can deny that there are entirely too many half-pint milkers around (and being sold to unsuspecting novices who have heard that goats give a gallon of milk a day). The reason is simply poor culling practice — often starting with the selection of the buck. Just as a good buck will improve many future milking does, a poor buck will drag them down. Cull a poor buck today and save yourself the trouble of culling dozens or hundreds of poor does in the future.

Now, it's true that showing, and in fact the entire registration system, sometimes aids and abets this. There are goats that place high in the show ring but don't produce enough milk to pay for their grain ration. It is true that too many goats are registered (and sold for high prices and allowed to produce more high-priced goats), simply because they are purebred. If you just want milk you can ignore the faults in the system, but don't throw out the baby with the bathwater. Ignore the faults, but not the entire system, if you want to produce milk efficiently.

That means following the system as it was meant to be followed. Breed the best to the best and cull the rest. *Cull* means destroy; it doesn't mean passing on substandard animals to the first sucker who comes along or selling them as pets only to have them or their offspring find their way back into someone's dairy. Too many people take the short-term view of the economic loss incurred, and as a result short-change themselves, and future goat raisers, for the long term.

Consider the Alternative

Your chances of improving your herd are practically nil if you breed your does to a neighbor's nondescript pet buck simply because he happens to be cheap and available, or if you buy a buck just because he's cheap. You aren't going to milk the buck, but never forget that you're going to milk his daughters, and hopefully many future generations. If he doesn't have the genetic potential for milk production, his daughters won't have it either. If the buck is not much better than the doe, you're not working for breed improvement. In fact, you're not even breeding goats: you're merely using the buck as a doe freshener, to get a little milk.

Actually, there are cases where goat keepers (as distinguished from goat breeders) are only interested in freshening does, because they only want milk, and don't even keep replacement animals. Since the buck has no effect on the lactations of the does he breeds, he would have no effect on the herd if all his offspring were routinely butchered. This seldom happens, however, and the downgraded goats are foisted upon a world that already has too low an opinion of these valuable animals.

Choosing a Buck

So how do you choose a buck that will produce superior offspring?

You start by examining his pedigree, the record of his ancestors. If it's milk you want, make sure there's milk in that pedigree. While there admittedly are lovely grades that make milk by the ton, there is no way of knowing who their ancestors were or how good they were. A pedigree and milk production records of several generations of forebears might not be insurance, but they're valuable management tools and much better than flying blind.

Next consider conformation. If you have a doe with poor udder attachment or weak pasterns or any other fault that might affect her productivity and usefulness, you certainly won't want to breed her to a buck whose dam and granddam had the same faults. On the contrary, you want a buck that is particularly strong in those areas, so his daughters will be better than their mothers.

Now, suppose you purchase a fine buck of impeccable breeding, excellent health, and ideal conformation. Are your problems over? Not quite.

Buck goats are much more masculine in appearance than does, as this Nubian demonstrates. Despite their powerful builds and disgusting (to humans) habits, bucks that have been well-treated and trained can be quite docile.

If you have four does, a fair average for a homestead herd, you can expect four doe kids the first year. Chances are you'll want to keep one or more of them. After all, didn't you buy the high-powered buck to improve your herd?

But then, it's evident that the next breeding season you'll be making father-daughter matings. This isn't necessarily bad. While many people attach the human-oriented stigma of incest to such matings, in the hands of an expert they are actually the surest and fastest way to breed improvement. But few beginning or backyard goat raisers are experts in animal genetics, if only because a herd of a few animals would require years and years to give the breeder the experience necessary to be an "expert." It's sufficient to say here that inbreeding emphasizes faults as well as good points; it's nothing to be dealt with haphazardly. (Actually, there is some evidence to suggest that inbreeding affects goats less than some other animals. But are your original goats good enough to be perpetuated — or should they be upgraded by outcrossing?)

So when your herd sire's first daughters come into heat you'll want another buck.

Minimizing Faults

This is a good place to note that no animal is "perfect." All have faults of one kind or another, to a greater or lesser degree. It's the job of the breeder to eliminate those faults as much as possible in future generations, while at the same time preventing new ones from showing up.

An illustration of this would be a doe with very good milk production but a pendulous udder. That udder fault is going to shorten her productive life, it will make her more liable to encounter udder injury and mastitis, and so forth. So you'll want to breed such a doe to a buck that tends to throw daughters with extra-nice udders, in hopes that the offspring will have both good production and acceptable udders. They won't be extra-nice in view of the genes contributed by the dam, but they can be improved by proper buck selection.

The problem here is that, with four different goats in a small backyard dairy, there are likely to be at least four different faults! It's unlikely that even a good buck will be strong enough in four different areas to compensate for all of them. From the standpoint of breed improvement then, each doe in your barn is likely to be best matched by a different buck.

These are real and important and practical considerations. But we must also mention some more commonly voiced objections to keeping bucks.

Other Pros and Cons of Keeping Bucks

Unlike does, bucks do smell, especially during breeding season. Girl goats (and some goat people) are inclined to like the aroma, but it will not only hang over your barnyard, it will pervade your clothes, and even your living-room furniture will get to smelling like ripe billy goat, which for most people is less than desirable. Your neighbors might also have an opinion on this.

Also of interest to people who are new to goats are what they often call the buck's "objectionable, disgusting habits." Most city people are shocked when they find out that the cute and playful buck kid grows — astoundingly rapidly — into a male beast who not only tongues urine streams from females (and makes funny faces afterward) but who also sprays his own beard and forelegs with his own urine. This is natural goat behavior, but be that as it may, even many broad-minded people find it difficult to accept gracefully. Needless to say, the lovable buck kid loses a few friends when he reaches this stage.

Bucks are powerful animals — I've seen them snap 2×6s just for kicks — and one that has not been raised properly or finds himself in an untenable position can be a dangerous animal. (I have never owned a buck that was any more hostile or aggressive toward humans than a doe. Still, they haven't been effeminate either, which would be a fault in a buck. But enough other people speak of "mean" or "dangerous" bucks, so it seems likely that they exist, and you should be forewarned.)

Because they are powerful, and because of their natural sexual instincts, a buck requires much more elaborate and expensive housing than the does, especially during the breeding season. They must be housed separately if only to avoid off-flavored milk. And an inadequately penned buck will soon be found with the girls.

In spite of all this, there are many practical and logical reasons for keeping bucks even for small herds, and many people do. Despite the arguments listed here, some people simply aren't interested in improving their herd, much less the breed. They just want milk, and perhaps a little meat. Most of these don't care to expend the time and effort it takes to learn about pedigrees or to study them. Increasingly, many aim to be self-sufficient in dairy products, and how self-sufficient can you be without a herd sire? If you start with fairly decent or above-average does, and get a buck that's as good or better, it certainly isn't a catastrophe. But you should at least be *aware* of the information we've discussed here. If you're going to be a goat breeder, be the best you can, within the parameters you have set for yourself.

While there are many advantages to buying a proven sire — a buck you know is not sterile and who throws daughters with the traits you want in your herd — such bucks are either very expensive or old and otherwise worn out. Most bucks are sold as kids, fresh off their dam's colostrum.

As mentioned, most good breeders dispose of buck kids at birth, even very good ones, because there is little demand for them. Half of all kids born are bucks, and only a small fraction of those are needed for breeding. Only the very best are kept, and almost invariably these have been reserved far in advance.

Caring for the Buck

Buck kids are raised very much like doe kids. They grow a little faster but take longer to fully mature. Yet, even though a buck may not stop growing until he's 3 years old, he is capable of breeding a doe by the time he's 3 or 4 months old. Don't let his size or his kidlike demeanor fool you! Separate penning is necessary almost from the beginning, at least by 2 months of age.

Housing

A buck's housing must be especially sturdy. It's difficult to "overbuild" a buck pen! The barn or shed can be simple, but it should have walls of 2×6s or 2×8s, or cement blocks. Walls, pens, and fences must be strong enough to withstand the assault of a 200-pound battering ram with a head as hard as a sledgehammer. A shed of 6 feet by 8 feet will provide sufficient shelter, with an exercise yard at least 6 feet by 30 feet. Posts — extremely sturdy and well set in the ground — should be no more than 8 feet apart.

The buck barn and yard should be sturdily constructed. This one is attractive and serviceable, but the herd sire would enjoy it more if it were less square. The ideal yard can be as narrow as 6 feet wide, but it should be at least 30 feet long. Another improvement would be a means of feeding and watering the buck without entering the pen.

Feeding

Feeding a buck is less complicated than feeding a doe, since the male needs less protein and minerals. They can get the same feed, with one exception: don't feed bucks pure alfalfa. It contains more calcium than a buck can use and can lead to kidney stones. Feed a good-quality grass hay.

Training

Early training pays off. Teach a buck kid to lead, with a halter, while he's still small enough for you to handle. Some people teach bucks to lead by the ear. Use patience and persistence and he'll remember, which can be very helpful when he weighs over 200 pounds and could easily over-power you.

Again, that size and strength can be dangerous when does are in heat. Any male of this size is to be handled with respect and caution.

When and How Often to Breed Your Buck

A buck can be used for limited service even before he's a year old. Most authorities say he should be limited to ten to twelve does his first year. A mature buck can service more than one hundred does a year, and of course, with artificial insemination many more than that.

Artificial Insemination

Artificial insemination (AI), by the way, is another option. A new goat owner isn't likely to make this a do-it-yourself project, but if anyone in your area is using AI and will breed your does this could eliminate your buck problems. Watching someone else is the best way to learn how to do it yourself. (See chapter 10.)

BREEDING

A goat obviously must be bred in order to produce kids — and milk. Those 150 or so days between breeding and kidding are extremely important to the goat, and for the first-time goat owner especially, they are anxious ones.

The Doe's Cycle

It all starts with the doe's estrous cycle or heat periods. Goats and most other animals "cycle"; that is, they are fertile only for relatively short periods at more or less regular intervals. Unlike cows, or hogs, or rabbits, which come into heat year-round, goats generally come into heat only in the fall and early winter. A doe will accept service from a buck only when she is in standing heat, usually. If she is not in heat, copulation won't result in pregnancy anyway because the sperm and the ova aren't in the right place at the right time.

Seasonal breeding has decided advantages for animals such as deer and goats in the wild. Their young are born when it isn't too cold; there is plenty of lush, milk-producing feed for the mother and tender grasses and leaves for the young to be weaned on; and the offspring are fairly strong and independent by the time the weather turns harsh again. Desirable as such an arrangement may be for wild animals, it puts the dairy goat farmer in a bind.

Lactation and Seasonal Curves

In chapter 2, we examined the lactation curve. If you plot such curves for several does, all of which have been bred at more or less the same time, it's apparent that the goat farmer will be swimming in milk during part of the year and dry as a bone in another part. This is perhaps the single most serious drawback to commercial goat dairying. If people want to buy goat milk they want to buy it regularly, not just when it's in season. (See appendix A on milking through for one potential solution to this problem.)

The normal lactation curve is reinforced by seasonal curves that are equally normal in both cows and goats, due to feeding conditions, weather, and other factors. Animals simply produce more milk in summer than in winter.

Add to that the fact that more people want goat milk in winter than in summer, and it's easy to see that the poor goat farmer has a problem.

Backyard dairy operators and homesteaders share the same dilemma, to some degree. If you have just one goat, even if she has a lactation of 10 months, you'll be without any milk at all for 2 months of the year. With two goats you can attempt to breed one in September and one in December. Then, theoretically, you will never be without milk, but a look at the lactation curves plotted together will show that your milk supply will be far from steady. You'll have too little or too much far more often than you'll have just enough.

The point here is that you'll want to have your does bred as far apart as possible, but while still avoiding the risk of having a doe miss being bred at all. With some does, in some years, even December may be pushing it; they simply won't come in heat again until September. There are kids born in every month of the year, but as a practical matter for the small-scale raiser, you can't count on out-of-season breedings.

Detecting Heat

For many beginners, and especially those with only one or two goats, it's very difficult to tell when a goat is in heat. The usual signs are increased tail-wagging; nervous bleating; a slightly swollen vulva, sometimes accompanied by a discharge; riding other goats or being ridden by them; and, sometimes, by lack of appetite and a drop in milk production.

If a buck is nearby there will be no doubt: she'll moon around the side of her yard near the buck pen like a lovesick teenager.

If you lack a buck and have trouble detecting heat periods, or just want to make very sure she's in heat before you make a several hour trip to a buck, you might use this trick. Rub down an aromatic buck with a cloth, or tie one around his neck for a couple of hours. Gingerly poke it into a canning jar and screw the lid on. When you suspect your doe is eager for male companionship give her a whiff of that cloth and your suspicions will quickly be confirmed or denied.

But you can't breed a doe with canned buck aroma. If you don't have a buck you'll have to take her to one.

Transporting the Doe

When we had neither a buck nor a pickup truck, we used to transport does in the trunk of the car. Some people were aghast at this dangerous practice. But since it always worked fine for us, and there's no way I can afford a livestock trailer, and a pickup truck (even with side racks) is far more hazardous for the goat, I still favor the car trunk. But you decide for yourself. (With the recent popularity of pickups with toppers and sport-utility vehicles, maybe this is no longer a concern.)

If your car isn't too fancy or if you really love goats, she can ride in the backseat. If she's lying down, she will not "disgrace herself," as one puritanical old goat book put it. Some goats tend to get carsick standing up, and will be too woozy to be interested in breeding after the trip. On the other hand, some enjoy riding in a car as much as dogs do, even to the point of sticking their heads out the window, which could easily cause an accident on a busy freeway among drivers who have never seen a goat riding in a car.

Artificial Insemination

There is another possibility that interests many people: artificial insemination, or AI. This is a relatively new development in the goat world. It's not as common as the artificial insemination of cows and it's not 100 percent effective, but it has great potential. It has gained wide acceptance in the past 20 years.

From the standpoint of breed improvement, there is nothing better. Anyone, anywhere, can use the finest bucks available, and at low cost. In many cases one straw of semen (a *straw* is the glass tube the semen is stored in) costs only a tenth of the same buck's standing stud fee, and you can use bucks that are thousands of miles away. Does can even be bred to bucks that are long dead; semen can be stored for years. Best of all, you can use *proven* sires, bucks that already have daughters, or even grand-daughters, that are already milking and have production records. And the inbreeding problems mentioned in chapter 9 are easily eliminated with AI. Even goat keepers who aren't overly concerned about inbreeding can readily see the advantages of not having to keep a buck or having to traipse all over the countryside with lovesick does in the car.

At the same time, AI isn't the final answer to every goat owner's breeding concerns. In many cases, it will be necessary for you to do the inseminating yourself. And although it's a relatively simple procedure, you won't learn how to do it by reading a book. Beginners typically have a 50 percent success rate. The best way to learn is by watching an experienced inseminator and asking plenty of questions. If you'll have to buy and maintain a liquid nitrogen semen storage tank, your breeding costs for just a few does will go up appreciably.

All of this is far beyond the basics of raising goats and the scope of this book. Look to the goat periodicals for ads for artificial insemination companies and contact them. Of course, you can also get the information you need to get started from any goat raisers in your area who are using AI.

Successful Breeding

A doe will be in standing heat for 24 hours, although this varies widely. If she is not bred, she will come into heat again in 21 days, although this too is an average that varies considerably.

If a doe is serviced and still comes back in heat, there could be several causes. She might not have been bred at the most opportune time. Maybe one more try will do the trick. (It isn't necessary to leave the buck and doe together for long periods. If the doe is really in standing heat, one service is sufficient, and that won't take more than a minute, which sometimes seems silly after you've spent an hour on the road and still have to drive home again!

Breeding Problems

Sterile bucks are rare, and if a buck is sterile obviously none of the does he serves will conceive. However, sperm can have reduced viability at certain times due to overuse of the buck and other factors.

If a doe simply will not get bred, the most common cause is cystic ovaries — a growth preventing ovulation — and she is worthless. Overly fat does are often difficult breeders because of a buildup of fat around the ovaries.

Another serious condition, although it's not as common as we once thought it was, is *hermaphroditism*, or bisexualism. For somewhat technical reasons the term "intersex" is now preferred in some circles as being more accurate but "hermaphrodite" is easier to use and more colorful.

The First Hermaphrodite

The term comes from Hermaphroditus, the son of the Greek gods Hermes and Aphrodite; his name is a combination of theirs. According to myth, the nymph Salmacis fell in love with Hermaphroditus, who was bathing in her fountain. She wanted the two of them to achieve complete union, and her wish was literally granted.

The hermaphrodite looks like a doe externally, but it actually has male organs internally. Not all have obvious external abnormalities. Carefully examine the vulvas of newborn kids. A growth about the size of a pea at the bottom of the vagina is abnormal. Unusual behavior in a normal-appearing doe kid is cause for suspicion. Intersex goats are often overly aggressive or unusually withdrawn.

In goats, the condition is often related to the mating of two naturally hornless animals. The genetics gets a little complicated, but basically you must determine whether a naturally hornless buck is homozygous or heterozygous; that is, whether or not it inherited a gene for horns from either of his parents. Theoretically, there can be no homozygous does because they'd be hermaphrodites and couldn't have offspring. Both types of bucks will produce some hermaphrodites when bred to hornless does, according to theory, but the homozygous hornless bucks will produce more.

If either the buck's sire or dam were horned, he's heterozygous. If neither parent was horned, you can't be sure what he is without seeing a number of his kids. If any of the kids have horns, the buck is heterozygous. If all the kids are hornless, even out of horned does, chances are the buck is homozygous.

All of this is of great interest to geneticists and large goat breeders and people who take a keen interest in breeding. But the average back-yarder or homesteader is better off to follow the lead of the major commercial goat farms and just avoid hornless-to-hornless matings.

When to Breed

Doelings are sexually mature as early as 3 or 4 months of age. They should not be bred at that stage of development. In most cases spring kids that are well developed and healthy should be bred when they weigh about 80 pounds and are 7 or 8 months old, which means they'll kid at 1 year of age. Breed by weight, not by age. Being bred too early will adversely affect their growth and milk production; being bred too late does not contribute to their health and welfare; it's expensive to keep dry yearlings; and records show that does that kid at 1 year of age produce more milk in a lifetime than those that are held over. Many people mistakenly hold back young does because "they look so small" or because 7 months seems so young. With proper nutrition, they'll produce healthy kids, and keep on growing themselves.

Drying Off

The older doe will be, or should be, still milking when she's bred. But advancing pregnancy will cause most does to dry off. Some people who really want milk will continue milking a naturally drying-off doe as long as she's giving a few squirts; others figure even a pound isn't worth their trouble, stop milking, and the doe dries up.

In any event it's a good practice to dry off a doe 2 months before her kids are due. Milk production makes great demands on a doe's body. So do growing unborn kids. The kids — and the health of the doe herself — are more important than the milk. In most cases simply quitting milking and reducing the grain ration will cause the animal to dry off naturally. In cases of extremely heavy or persistent milkers it may be necessary to milk her out at intervals. Some people milk once a day for a few days, then every other day, then stop. This sends mixed signals to the brain. It's better to just stop, period. Reducing or eliminating grain will help an animal dry off. (See the box below and appendix A.)

How to Dry Off a Doe

The best way to dry off a doe is to stop milking her.

1. At the last milking, use a dry-cow antibiotic infusion to reduce the possibility of mastitis. (A teat is infused by injecting an antibiotic into the teat canal.) Also use a teat dip. (See chapter 8.)
2. Switch to a dry doe ration, which has reduced concentrates and a lower protein content.
3. Dip the teats twice a day for the first 4 to 5 days. Pressure builds in the udder and signals the body to stop producing milk.
4. If there is still substantial pressure in the udder after 4 days, milk out the doe, reinfuse the udder, and start the procedure over again.

Merely decreasing the frequency of milking — perhaps to once a day, then every other day — can lead to fibrosis of the udder and lower milk production in future lactations.

Feeding the Pregnant Doe

No good dairy animal, cow or goat, can eat enough during lactation to support herself and her production. That's why she requires a rest to build up her body before peeling off her own reserves to fill your milk pail. It's been said that for each pound of increase in body condition during the dry period, a Holstein cow will produce an extra 25 pounds of milk, a Guernsey an extra 20 pounds, and a Jersey an extra 15. We could anticipate proportional results with goats.

This is not to suggest that a pregnant animal should be overconditioned or fat. A dairy animal should never be fat, just in good condition. In fact, fat causes problems in breeding and pregnancy. But the goat needs a well-balanced diet. Too much feed produces kids that are too large to be easily delivered. Excess minerals in the doe's diet produces kids with too-solid bones, which also causes difficulty. A fibrous diet with rather low protein is ideal for the first 3 months of pregnancy when the kids are developing slowly.

Most of the kids' growth comes in the last 8 weeks of pregnancy. During this period the ration should be changed gradually, not only because two-thirds of the kids' growth is taking place, but because this is when the doe needs to build up her own reserves for her next lactation. High protein still isn't required — about 12 percent will do — but there is definite need for minerals and vitamins, especially iodine, calcium, and vitamins A and D. Bulk such as is provided by beet pulp or bran is required, and molasses will supply some iron as well as the sugar that helps prevent ketosis, and it has a desirable slight laxative effect.

Finally everything is ready. The goat stork cometh.

KIDDING

The "miracle of birth" is aptly named. Like all miracles it's invested with wonder, awe, excitement, and joy. There have been cases of people who would have nothing to do with goats — until they saw newborn kids frolicking in fresh clean straw and fell in love. (I'm married to one of these people.)

There is little doubt that the first kidding season brings the new goatkeeper excitement that is hard to duplicate in today's electronic and plastic world. Most of them, judging from the mail I get and my own first experiences, are scared silly as parturition approaches. Most of this fear comes, I believe, from reading books and articles describing all the things that can go wrong. You expect the worst. But goats have been having kids all by themselves for thousands of years. While problems are possible, 95 percent of all goat births are completely normal and won't even require your assistance. The chances for a normal birth are enhanced by proper feeding and management, especially during the latter stages of pregnancy.

Within minutes of entering the world, the newborn kid will struggle to its feet and search for its first meal.

Kidding Supplies to Have on Hand

- Iodine and dipper
- Clean cloth towels
- Small clean water pail
- Feeding tube (lamb size) and syringe

- Alfalfa hay
- Nipples and bottles
- Colostrom (frozen)

Anticipating the Delivery

The average gestation period for goats is 145 to 155 days. There is a tendency for does with triplets to kid slightly earlier than does with single kids, but both are usually within this time frame. Some experts say there is evidence that goats and sheep can control the time of birth to coincide with copacetic weather conditions. Other people say they control it, all right, but usually to have the kids and lambs arrive on the coldest, most miserable night of the year. Either way, many goats seem to kid at about the same time with every freshening. Record that time for each doe and next year you might be forewarned.

Checking for the Signs

Start checking your animals frequently and carefully 140 days after breeding. When she is getting ready to kid, the doe will become nervous and will appear hollow in the flank and on either side of the tail. There may be a discharge of mucus, but this can appear several days before kidding. When a more opaque, yellow, gelatinous discharge begins it's for real.

Kids can be felt on the right side of the doe. It's good practice to feel for them at least twice a day. As long as you can feel them they won't be born for at least 12 hours. Also if you feel the doe regularly you'll be able to notice the tensing of the womb. After this, one of the kids is forced up into the neck of the womb, causing the bulge in the right side to subside somewhat. This will be noticeable only if you have paid close attention to the doe in the days and weeks before. The movement of the first kid also causes the slope of the rump into a more horizontal position. At this point you can expect the first kid within a couple of hours.

Many people look to "bagging up" or enlarging of the udder as a sign of approaching parturition. This is unreliable. Some goats don't bag up until after kidding, and others will have a heavy milk flow far in advance. In some cases, if the udder becomes hard and tight, it might be necessary to milk out the animal even before kidding.

A better indication is a softening of the ligaments from the tailhead to the pinbone. Check these earlier so you know how they feel "normally." When they seem to have disappeared, the doe should kid within 12 hours.

Preparing the Facilities

Although humans try to take good care of their animals, we often complicate things. We've mentioned feeding of the pregnant goat, which can affect the ease with which she delivers. A free-ranging, experienced goat knows what to eat, but if we bring her food she must depend on our judgment. Likewise, the goat kidding outside on her native mountain range knows what to do when her time approaches. She is probably safer and in more hygienic surroundings on her mountain than in your barn. It's just about impossible to duplicate natural conditions for domestic animals.

There are innumerable instances of goat owners going to the barn for morning chores and finding a couple of dried-off, vigorous, and playful kids in the pen with their mother. But it's definitely preferable to have some idea that the kids are on the way and to make certain preparations for them.

The doe should have an individual stall for kidding. It should be as antiseptic as possible and well-bedded with fresh, clean litter. Something softer than long straw, such as chopped straw, is preferable if you have it.

Don't leave a water bucket in the kidding pen. It can be dangerous, and the mother isn't interested in drinking at this point anyway.

The Birth

One study of owners' records shows that 95 percent of goat births are uneventful. Your assistance will be unnecessary and perhaps even unwanted. During labor you'll have nothing to do: you don't even have to boil water, unless you want to make a cup of tea.

The Three Stages of Kidding

Kidding normally occurs in three stages. The first stage is when contractions of the uterus force the placenta, fetus, and fluids against the cervix, dilating it. This can last up to 12 hours in first fresheners, but older does are usually faster. Second-stage labor involves "straining," or contraction of abdominal muscles. This typically lasts about 2 hours or less, and ends in the expulsion of the last kid. The third stage involves the expulsion of the placenta, or *afterbirth*. Four hours is considered the norm for this, although it can take longer. But much longer and it's a *retained placenta*, which calls for the services of a veterinarian.

Assisting the Delivery

If the doe has been struggling for a while, seems concerned, and nothing is happening, you'll have to help. Insert your disinfected, lubricated hand and arm into the birth canal to find out what's wrong. A germicidal soap will serve as disinfectant; mineral oil or K-Y jelly are lubricants.

If you've never seen a newborn kid, this is not only scary, it's difficult to imagine what you're feeling for. Try to sort out the heads and legs, and if necessary rearrange them in the proper presentation position. In most cases, it will be a simple matter to "lead" the first one out the next time the doe strains. Pull, but very gently, working with the doe; otherwise, hemorrhage might result. Chances are the others will come by themselves soon after.

A *pessary*, available from your veterinarian, should be inserted into the vagina after manual exploration to minimize the danger of infection.

Goats usually have two kids, but three, four, and even five aren't all that uncommon. If no more come within half an hour and the mother seems relaxed and comfortable, you can assume that's all there are.

If you happen to encounter a difficult birth, a doe struggling to expel a dead kid for instance, get a veterinarian or experienced neighbor to help. But again, such help is seldom needed. Be aware of what can go wrong, but please don't make yourself sick worrying about it beforehand. Being anxious about an imminent birth is normal, but remember that ninety-five out of one hundred births are normal, too. Relish the experience!

Kid Presentations

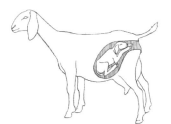

Normal. The kid's nose is positioned between the front feet. Picture it in your mind's eye and you'll see that this presents a more or less cone or wedge shape that gradually distends the vagina and makes the birth easiest.

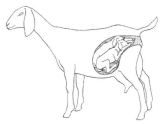

Multiple Births. Twins are more common than single births, and triplets, or even quadruplets, are not rare. In some multiple births, the kids and umbilical cords get all tangled up inside the womb.

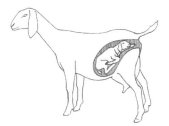

Foreleg Back. One leg is bent back, the wedge shape is missing, and the bent leg might complicate the birth. If the doe is having difficulty, it might be necessary for the caretaker to reach in and bring the bent leg forward.

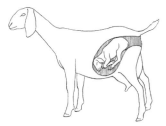

Breech. A breech delivery inverts the wedge shape, and can cause a difficult birth. The biggest danger is that if the sac breaks before the kid is completely out, it could suffocate.

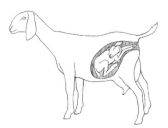

Head Turned Back. If the kid's head is turned back it might have difficulty entering the birth canal. The solution is to reach in and move the head so the nose is between the two front feet.

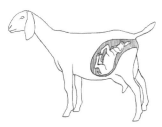

Upside Down with One Leg Back. In this unusual position, the kid is both upside down and has one front leg back. If the doe is having difficulty, the caretaker will have to go in and rearrange the kid for a normal presentation.

The Umbilical Cord and Afterbirth

In most cases the umbilical cord breaks by itself, or the doe severs it. If necessary, tie it off with a soft string (a shoelace is fine in an emergency) about 2 or 3 inches from the kid's body, and snip it off on the doe's side of the knot with sharp scissors.

One of the most important accessories of the midwife is a good supply of clean cloths, and lots of them! Wipe the kids off, paying special attention to make sure the mouth and nostrils are clear of mucus. The doe will gladly help by licking her newborns. If you are on a caprine arthritis encephalitis (CAE) or CL prevention program, of course, it will be up to you to make sure that the doe does *not* lick the kids (see chapter 8).

The umbilical cord then must be disinfected. Iodine spray is convenient, but better protection can be had by pouring some iodine into a small container, pressing it up to the kid's navel, then briefly tipping the kid over to ensure good coverage of the entire navel with the iodine. (At last! A use for those plastic 35-mm film canisters!)

Watch for the afterbirth. If the doe doesn't expel it within several hours, call a veterinarian. A retained afterbirth is nothing to mess around with. If it's just hanging out of the doe, don't pull on it. You might cause hemorrhaging.

When the afterbirth is expelled, dispose of it. Some does eat it, a natural instinct most wild animals have developed to protect their young from predators that might otherwise be attracted. Most goat keepers don't care for this little ritual, but it won't hurt the goat. But it won't do her any special good though, either, and she could choke.

Attending to the Doe

Put the kids in a clean, dry, draft-free place (a large box works well), and turn your attention to the doe. She has lost a tremendous amount of heat, even in warm weather. Offer her a bucket of warm water. Some does seem to appreciate the water even more if a cup of cider vinegar is added.

Most pet-type goats also get a special treat at this point; perhaps a small portion of warm bran mash or oatmeal or a handful of raisins. Provide some of the best hay you have — she's earned it.

Caring for the Newborns

The new kids need your attention just after they are born. Following are your responsibilities.

Kid Inspection

Examine the kids. Many people who don't want to bother raising buck kids for meat euthanize them at birth. (The easiest way is to drown them in a bucket of water.) And remember, only the very best bucks from outstanding dams and sires should be kept for breeding. One mature buck can breed a hundred does a year, and many bucks are kept for 5 years or more. Mathematically, this means that less than one in five hundred is needed as a sire. It's highly unlikely to be the one just born in your barn.

On doe kids, check for supernumerary (extra) teats. There are several variations of this condition, some of which make the animal worthless as a milker. If the extra teat is sufficiently separated from the main two, it might not interfere with milking and can even be removed at birth with surgical scissors or by tying a fine thread very tightly around the base and letting it atrophy. Actual double teats make the animal worthless. Bucks can also have double teats, and such animals should not be used for breeding.

If the tip of the vulva on doe kids has a pealike growth, the animal is a hermaphrodite (see page 157) and will not breed. It should be destroyed, and the mating that produced it should not be repeated. Not all hermaphrodites, however, will display this growth.

Unwanted or *cull* kids, bucks and does, can be raised for meat for your own table, or for sale. If you have a better use for the milk and don't want to bother with milk replacer and the work of raising kids, they can be butchered at birth and dressed like rabbits. Wait until they dry off, then they're easier to handle. (See chapter 15 for more on butchering and meat.)

Keeping the Kids Warm

Perhaps one of the most common kidding problems most people encounter is entering the barn on a blustery morning in late winter or early spring to find a newborn kid cold and shivering. If it seems to be doing all right, don't feel sorry for it and bring it in out of the cold. Place it in an enclosed box or pen padded with an old blanket or feed sacks, away from any hint of a draft, and with a heat lamp if the weather is really nasty. But don't let it get hot. A switch back to normal temperatures will be as dangerous as the cold that brought on the problem in the first place.

In the case of a severely chilled kid on the brink of death, more drastic action will be required. If you find one still wet and thoroughly chilled and nearly lifeless, one way to save it is to submerge it up to the nose in a bucket of water at around 105°F, which is about the temperature of the environment it just came from. When it has revived, dry it thoroughly, wrap it in a feed sack or blanket, put it in a box in a protected place, and watch it carefully.

However, such a kid might also be suffering from hypoglycemia, or low blood sugar. As the glucose level falls, the kid shivers, arches its back, its hair stands on end, and it moves stiffly. Eventually it lies down, curls up, becomes comatose, and dies.

This is a cutaway view of a warming box. Note the hinges at the top, on the wall. Such a box can keep kids more comfortable in extremely cold or drafty weather, but newborn goats can stand surprisingly low temperatures if they're in a draft-free place.

The remedy is to warm the kid and administer at least 25 milligrams of 5 percent glucose solution with a small rubber stomach tube. When the kid is showing signs of reviving, get 2 ounces of colostrum into it, with the stomach tube if necessary. Return it to the barn as soon as it's active.

If you do end up taking a kid into the house for any reason during cold weather (and almost everyone does), you're stuck with a goat in your house until the weather warms up. Even then you should harden it off by degrees rather than exposing it to the cold all at once.

How to Use a Stomach Tube

A kid that is too weak or comatose to suck must be fed with a stomach tube. (Stomach tubes are available from sheep- and lamb-supply houses, though a sterile catheter from a hospital or nursing home will work.) The stomach tube is a small, flexible plastic tube to which you attach a 60-mL syringe.

1. Slowly and gently push the tube down the kid's throat. Often, the kid will swallow the tube as you advance it, which helps.
2. When the end of the tube reaches the stomach, attach the syringe of colostrum or milk to the upper end of the tube.
3. Depress the syringe's plunger slowly to discharge nourishment directly into the stomach.

Note: Be absolutely certain that the tube is in the stomach *before* administering nourishment, so the liquid isn't forced into the lungs.

Colostrum

The kids will be standing and trying to walk on wobbly legs, perhaps within minutes of birth, and they'll soon be looking for their first meal. This must be their dam's *colostrum,* or "first milk," a thick, sticky, yellow, nonfoaming milk. Colostrum contains important antibodies and vitamins, and the survival of any newborn mammal is in jeopardy without it. It's so important, in fact, that if you buy a goat just a short time before kidding and she has no opportunity to build up antibodies to your particular location and therefore can't pass that protection on to her kids through colostrum, you might possibly encounter problems with sickly kids.

You can allow kids to nurse, or you can milk out the doe and feed the kids from a pan or bottle. Milking the doe is preferred, not only because of CAE and Johne's disease, and because it's important to start the kid on pan or bottle feeding from the beginning, but also because it's the only way you can be certain that the doe has colostrum and the kids are getting enough — and not too much, which can cause overeating disease. Either way, it's extremely important to get some colostrum into them within half an hour or so of birth. Put it into a kitchen-clean pan (as opposed to just a "barn-clean" one; stainless steel is ideal), and dip the kid's nose into it. It should be about 110°F. If the kid won't drink, check the temperature. It should feel warm, but not hot. A normal first feeding is about 8 ounces. *Warning: Never* heat colostrum in a microwave, and *never* boil it, or you'll destroy the antibodies. (See chapter 12 for more information.)

If for some reason the doe has no colostrum, you can make an emergency substitute.

Rule of Thumb

Ensure that each newborn kid will receive 1 ounce of colostrum per pound of body weight, three times daily.

Colostrum for Humans?

Most goat owners feed all the colostrum their does produce to the kids, or freeze any extra for emergencies, such as when kids are orphaned, or a doe doesn't produce enough. But if enough is available, it can be used as human food.

Called "beestings" in England, 1 cup of colostrum can replace 2 eggs and a scant cup of milk in baked goods. A custard can be made by mixing 2 cups of colostrum with 2 cups of milk and about ¼ cup of honey. Bake at 300°F until set, about 45 to 60 minutes. Sprinkle with nutmeg if desired.

Colostrum Substitute

3 cups milk
1 beaten egg
1 teaspoon cod liver oil
1 tablespoon sugar

Mix well. This, of course, lacks the maternal antibodies that provide immunity from common diseases of the newborn.

Commercial products are also available. Authors Mary Smith and David Sherman say in *Goat Medicine* that "numerous anecdotal reports suggest a marked increase in the incidence of neonatal septicemia" when such products are used. (Neonatal septicemia is usually associated with navel infection.) Because of this, the authors warn that colostrum substitutes "must be viewed skeptically." Be aware then, that there is no substitute for the real thing.

Another alternative is cow colostrum. This is generally frowned upon today because of the danger of Johne's disease and other health problems that can be transmitted from cows to goats.

If you are on a CAE prevention program, colostrum must be heat-treated. Although commonly referred to as pasteurization, that's not technically correct since temperatures required for pasteurization would turn colostrum into a pudding. Colostrum must be heated (preferably in a double boiler) to at least 135°F but no more than 140°F and held at that temperature for 1 full hour. (See box on page 172.)

While generally considered not fit for human consumption, there is a custard made from colostrum that some people consider a great delicacy (see page 170). Most goat owners won't have enough colostrum to worry about making use of it. If you have extra colostrum, freeze it for future emergencies or, in some places, possible sale. Pour it into an ice cube tray, and when it's solid, turn the cubes out into a plastic bag. One cube is just right for one feeding.

Heat-Treating Colostrum

Heat-treating colostrum and milk for kids has become common since CAE became a problem. But note that these forms of milk require two different procedures.

Milk is pasteurized by heating to 165°F and maintaining that temperature for at least 15 seconds. Alternatively, it can be heated to 143°F for 30 minutes. This is the method used in processing plants, but it's more trouble for the home dairy.

Colostrum, however, is altogether different. If it gets too hot it becomes a pudding. To pasteurize colostrum, heat it to at least 135°F but no more than 140°F and keep it at that temperature for 1 full hour. One simple way to accomplish this is to bring it up to the proper temperature in a double boiler or water bath (to avoid scorching). Then pour it into a preheated thermos and let it stand for an hour. Be sure you have an accurate thermometer and a quality thermos that will actually maintain that temperature for that period.

Cool the colostrum to about 110°F before feeding. (The ideal feeding temperature is about 103°F, but it will cool to that by the time the kids start drinking.)

RAISING KIDS

The excitement of freshening over, your goat barn can settle into a routine. For the first 3 or 4 days your doe will produce colostrum, the thick yellow milk so necessary for the kid's well-being. Then, there hopefully will be enough milk for both the kids and you. (After perhaps months of anticipation, what a treat!) And after 2 months or more of relative inactivity, your goat barn will be a hectic place. In addition to the usual feeding and cleaning tasks, you'll be milking twice a day — and raising kids.

Raising kids requires some knowledge and a lot of work and time. Goat raisers have many different opinions about how the job should be done, but none can deny that the first year of the goat's life, along with her breeding and prenatal care, is an important determinant in how she will behave and produce later.

The antics of young kids frolicking are so entertaining that most people can watch them for hours!

Early Feeding

If the doe has a congested udder or a very hard udder the condition often can be helped by letting the kids nurse for the first few days. The suggested procedure is to bring the kids to their dam every few hours, rather than leaving them together. While this entails more work, it eliminates a lot of commotion and consternation later on. First fresheners, which often have very small teats, are also frequently left with their kids if the milker's hands are too large. The teats will enlarge with time.

If you milk the doe, do it within half an hour of kidding and offer the kids some colostrum. It should be close to goat body temperature, which averages around 103°F. Colostrum scorches easily: use a water bath or double boiler to warm it (see chapter 11).

How will the kids be fed? Nursing is certainly the easiest method, but not necessarily the best, so far as goat breeders are concerned. Some people say it ruins the dam's udder, which is important not only if you intend to show her but also if you want her to have a long and productive life as a milker. Possibly a more important consideration is that you don't know how much the doe is producing or how much the kids are getting. Also, kids left with their mothers are much wilder than hand-raised kids. Another important consideration is that once a kid learns to suck its dam it will be difficult — maybe impossible — to teach it not to. Some does wean their kids relatively early, but there have been other cases where yearlings are still sucking, and your milk supply for the kitchen is lost. The only solution in those cases is complete separation. It's better to do it right away.

And then there is caprine arthritis encephalitis (CAE).

Pan Feeding or Bottle Feeding?

Kids not left with their dams can be pan-fed or bottle-fed.

Many breeders prefer bottle feeding because it's more "natural." They point out that with pan feeding the animal is forced to lower its head to drink and milk can get into the rumen, where it doesn't belong. Digestive upsets can result. Of more immediate practical concern, pans or dishes can get stepped in or tipped over, and Nubian kids end up with milk-sopped ears, which can result in skin irritation.

Still, pans are much easier to fill, wash, and sterilize than bottles and nipples. You won't need lamb nipples or bottle brushes.

A variation on bottle feeding that's very popular where large numbers of kids are fed is a large container (such as a 5-gallon pail) with special nipples. The nipples are attached to plastic tubes that reach to the bottom of the container. The kids suck on the nipples and the milk is drawn up through the tubes, just like drinking through a straw. Commercial units go by such names as Lam-Bar and Lamb-Saver. You can purchase the nipples and tubes and make one from a 5-gallon bucket.

Several brands of nipple buckets like this are available commercially, or you can make your own. The special nipples are attached to plastic tubes. Kids soon learn to draw milk up the tubes, like sucking on a soda straw.

If you're feeding more than a few kids with bottles, bottle racks are handy. You can feed as many kids as you want to at one time, without holding the bottles in your hands. You can buy plastic bottles with lamb nipples and wire racks that attach to board fencing, or you can easily construct a rack to hold soda or beer bottles. (See page 176.)

Whichever feeding method you decide to use, pan, bottle, or nursing, once you start it you're stuck with it. It's difficult to teach a kid to drink from a bottle once it's used to a pan, and the other way around.

Likewise, after the first feedings of colostrum, milk may be fed warm or cold, but be consistent to avoid digestive upsets. Feeding on a regular schedule is important for the same reason. Most people feed warm milk.

Cleaning Kid Bottles

Kid bottles can be difficult to clean with a brush. To clean hard-to-reach places, place about 1 tablespoon of clean coarse sand or aquarium gravel in the bottle with a drop of detergent and a little water, and shake vigorously. The sand or gravel can be strained, rinsed, and reused.

Make Your Own Bottle Rack

In a 1 x 4 of any length you want, cut or drill holes large enough to admit the necks of the bottles you'll be using. Nail the 1 x 4 at a right angle to a 1 x 10 or 1 x 12 to form an L; this allows the bottles to rest on the larger board with just their necks and the nipples poking through the holes. Using a 1 x 6 board for backing, fasten the rack to a fence at about a 45-degree angle and at a convenient height, depending on the size of the kids. Finish sides with a piece of ½-inch plywood as shown below.

Hungry kids butt udders (and bottles), and when they get large enough, you'll probably need another device to hold the bottles more firmly in place. And occasionally a kid will pull a nipple off a bottle, dumping the milk. But overall we've found this rack to save a great deal of time and labor.

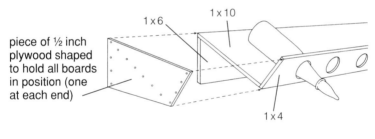

piece of ½ inch plywood shaped to hold all boards in position (one at each end)

1 x 6
1 x 10
1 x 4

Frequency

Everyone who has raised a few kids seems to have an opinion about the best method, in terms of nursing, pans, bottles, frequency, amounts, and weaning time. The fact that their kids survive and thrive is evidence that all of these methods work, in any combination. What it comes down to, then, is a matter of personal preference and convenience.

As with any baby, frequent small feedings are better than infrequent large feedings. Some people feed their kids every few hours. But if you have a job or a busy schedule, this might not be possible. In that case, feeding every 12 hours will be more convenient. Either method works, as long as you're consistent.

Of course, there are limits and guidelines. If at all possible, the colostrum feedings of the first 3 days should be 8 or even 6 hours apart,

Moderation Is Key

This is one area where the old saying, "The eye of the master fatteneth the cattle," is especially true. You obviously don't want to starve the kids. But don't kill them with kindness either. This happens most frequently by overfeeding milk and causing scours, which can be fatal. You don't want kids to be "fat and healthy" because fat *isn't* healthy for a dairy animal. Strive for condition, not overcondition. The kid should be producing bone, not fat, to develop her full potential in later life.

meaning three or four times a day. From then until the kids are 1 week old (the next 4 days), feeding anywhere from 4 to 8 ounces three times a day is recommended. During their second week, they will probably work up to 12 ounces, in three feedings 8 hours apart. And from then on, you can expect to feed about 1 quart of milk to each kid, twice a day.

But none of this is rigid or a matter of life or death. Some kids simply won't want this much. If they don't act sickly, they're okay. Some people give the kids all they will drink, and that's okay too, as long as the kids don't scour. (If they do, see the advice on scours in chapter 8, which includes limiting milk intake.)

If you feel lost without specific directions to follow, these will help. But remember that observing your animals is the real key to good husbandry! If they act hungry, check the amount you're providing and their body weight to make sure you're feeding them enough. If they don't want all you offer, check to make sure they're just full, not sick. This approach is far better than saying a kid *must* consume a certain amount each day.

As a rule of thumb, expect a newborn kid to consume about 2 cups of colostrum a day, in three or four evenly spaced feedings. This gradually increases to roughly 2 or 3 pints of milk or milk replacer a day at weaning, if they're weaned at 8 to 10 weeks. Offer warm water after each feeding of milk.

It's best to leave them a little on the hungry side, as this will encourage the consumption of solid foods, which will help develop the rumen. They will start nibbling on fine-stemmed hay in a week or so and on grain (18% kid starter) soon after. The more solids they consume, the less milk they'll drink.

Suggested Feeding Schedule for Kids

AGE	FEED	AMOUNT	FREQUENCY
Birth–3 days	Colostrum	12–14 oz/day	3–4 times/day
4–7 days	Goat milk	12–24 oz/day	2–4 times/day
1 week	Milk or milk replacer	36 oz	2–4 times/day
2 weeks	Milk or milk replacer	32 oz	2–3 times/day
	Good hay	Free choice	N/A
	Water	Free choice	N/A
3–8 weeks	Milk or milk replacer	32 oz	2 times/day
	Good hay	Free choice	N/A
	18% starter grain	As much as will be cleaned up in about 15 minutes	2 times/day
	Water	Free choice	NA

N/A = not applicable.
Note: These amounts are approximate and guidelines only. Base actual amounts on the appetite and condition of your kids.

One way to judge your relative success is to weigh each kid once a month. Ideally, they should gain about 10 pounds per month, for the first 5 months. For example, a kid weighing 8 pounds at birth would then weigh about 38 pounds at 3 months of age.

Weaning is another controversial area. On average, it's a safe guess to say that most kids are completely off milk by 8 to 10 weeks of age.

Still, some people feed milk for much longer, as long as 6 months, in some cases. This certainly isn't necessary, and according to Dr. Leonard Krook of Cornell University it may actually be harmful. Kids overfed calcium (milk is high in calcium) are likely to develop bone troubles in later life. In addition, we want that rumen to develop. That requires hay.

Milk Replacer

If you want all the milk your goat produces for yourself, the kids can be fed milk replacer. However, be sure to use milk replacer made for sheep or goats, not cows. Calf milk replacer is not high enough in fat, and goats will not do well on it. In fact, a large commercial dairy that has tried several brands of milk replacers and keeps meticulous records on all its goats, claims that even when the kids looked fine while on replacer, 2 years later most of them were dead or had been culled because they lacked stamina. Yet, many goats are raised on milk replacers. There are very few hard-and-fast rules about anything connected with kid raising.

Weaning

Early development of the rumen is extremely important for later production. Most kids will start to nibble at fine hay by the time they're 1 week old. They should be encouraged to do so with kid-size mangers and frequent feedings of fresh, leafy hay. Hay or forage is more important than grain. Encourage that.

This also translates into limit-feeding milk. When milk consumption is limited to 2 pints a day, consumption of dry feed is encouraged. This increases body capacity, with a corresponding increase in feed intake and digestion. Research has shown that at 2 months of age a weaned kid has a reticuloruminal capacity five times as large as a suckling kid of the same age.

Wean by weight, not by age. The usual goal is two to two and a half times the birth weight. The primary consideration should be whether the kid is consuming enough forage and concentrate to continue to grow and develop without milk.

At weaning, most breeders feed a calf starter ration: ½ pound, twice a day. The kids should always have access to good hay. At 6 months they are switched to a milking ration. By 7 months doelings weigh 75 to 80 pounds and are bred. Milk-fed kids weighing 20 to 30 pounds are in great demand as meat in some localities, especially at Easter and Passover.

Wean by Weight

Kids can be weaned by age or by weight. While it's possible to wean kids as young as 4 weeks old, 8 weeks is considered optimal. Later weaning costs more in milk and labor, and it retards rumen development. But weaning by weight is better because it prevents unhealthy or undernourished kids from being weaned too early and can reduce weaning stress. Studies suggest that weaning at a total body weight that is 2.5 times the birth weight produces good results. This is usually about 19 to 22 pounds.

Castration

If bucks will be slaughtered at a few months of age, castration is of little, or even negative, value. Intact males grow faster because of the absence of the stress caused by castration. If they will be held longer, they should be castrated before weaning to avoid a "buck odor" in the meat. Buck kids kept for meat require no special diet, but some chevon (goat meat) aficionados claim milk and browse produce the best meat. Butcher kids don't need a lot of grain. It has little effect on the palatability of the meat because the fat is deposited in the kidney and pelvic regions rather than in the muscle, and too much grain can lead to urinary calculi and other problems.

No matter which method of castration is used, a tetanus vaccine is required. This is usually given at 3 to 4 weeks of age, along with a vaccination for *Clostridium perfringens* types C and D, the bacteria that cause enterotoxemia.

Until they are about 1 month old, bucks can be surgically castrated without anesthesia. They can also be neutered with a small Burdizzo emasculator. After about 1 month, the surgical procedure should be used and with an anesthetic. A chemical method is also available.

Surgical Castration

A sharp, sterilized scalpel and an assistant are needed for this procedure.

A helper holds the kid by the hind legs, his back to the helper's chest. A quick, clean incision is made with the knife, and the testicle is grasped and pulled out. The other testicle is likewise removed, and the wounds sprayed with antiseptic.

Since this is a surgical procedure, it's best left to a trained veterinarian. In fact, in England it's the law that goats over 2 months old can be castrated only by veterinary practitioners, using anesthesia. Obviously, this is nothing to be taken lightly by an amateur who lacks knowledge, training, and experience.

Undesirable Alternative

It's also possible to castrate with strong, tight rubber bands made for the purpose and applied with a special tool. The special bands are slipped over the scrotum above the testes. Once the bands are applied, the testicles atrophy and the bands and sac fall off in a few weeks. Similar to dehorning with rubber bands, this has always been the least-recommended method of castration and is rapidly falling out of favor.

Burdizzo

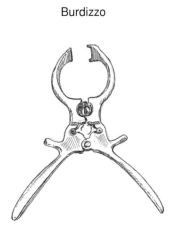

The Burdizzo, or emasculator, is used to crush the testicle cords to neuter young males. These are available from farm-supply stores and catalogs.

Elastrator, open (left) and closed (right)

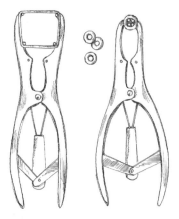

The Elastrator uses special, small, very strong rubber bands to neuter buck kids. This tool can also be used for dehorning. Easy to use and once quite popular, many people now consider it inhumane.

13

MILKING

The new dairy-goat raiser must learn about goat feeds and nutrition, about bucks and breeding and raising kids, all for one purpose: to get milk. Milking, therefore, is at the apex of the pyramid of all goat-keeping skills.

Most goats are milked at regular 12-hour intervals and according to a regular routine. Milking at 6 A.M. one day and 9 A.M. the next is one of the easiest ways to depress milk production. You might milk at 7 A.M. and 7 P.M. or at noon and midnight, but it should always be as close to 12 hours apart as usual and always at the same time. (See appendix A for an explanation of this.)

This has been the conventional wisdom, which most people follow. However, recent studies in France indicate that milk production in goats isn't affected if other schedules are followed, such as milking 10 and 14 hours apart, or even 8 and 16 hours apart, as long as the same schedule is followed every day. Whatever your schedule, once you become a goat milker your daily routine will be set not by your favorite television show, but by the goats.

Of even more importance than regularity is sanitation. One of the main reasons for keeping goats is having milk better than any to be found in the supermarket dairy case. This requires not only a knowledge of dairy sanitation but a rigid adherence to sanitation principles.

A milking stand like this makes that pleasant twice-a-day chore even easier and more enjoyable. Goats quickly learn to jump onto the stand at milking time.

Milking Essentials

What do you need to make milking comfortable and efficient? Equipment and good sanitation.

Equipment

Milking equipment can be simple or elaborate. You could, if you wanted to, milk into a bowl from the kitchen cupboard, make a milk strainer from two inexpensive funnels, and store the milk in the refrigerator in fruit jars. At the opposite extreme, you could spend hundreds or even thousands of dollars on milking equipment.

If you intend to milk 730 times a year for the foreseeable future you will get more satisfaction and better quality from proper equipment.

Milking Pail

A 4-quart seamless stainless steel half-moon hooded milking pail is no luxury, despite its fairly high price. Many of these have been in daily use for 20 years or more, which brings the per-use cost down to a pittance. Being seamless and stainless, these pails are easy to clean and disinfect. The hood and handle are removable to enable thorough cleaning. They're made especially for goats, naturally, so you'll have to check the goat-supply houses currently advertising in goat publications to find one. This is the one piece of goat equipment I couldn't do without.

Whatever you use, avoid plastic. No amount of cleaning can get the bacteria out of the pores in plastic, and you'll soon end up with a product that's unfit for human consumption.

Some people refuse to use any dairy equipment made of aluminum. This might be due to some misconceptions. It's true that aluminum isn't allowed in grade A dairies, but that's because their equipment must endure a great deal of very vigorous cleaning. The aluminum gets scratches, resulting in the problems associated with plastic. Some are no doubt concerned about reported connections between aluminum cookware and human health, especially Alzheimer's disease. But you don't cook milk in your milk pail or strainer.

Aluminum home dairy equipment is *anodized*, or coated with a protective film. Milk never touches the aluminum unless the coating is removed by abrasive cleaning. The nicks and scratches that lead to potential problems in the commercial dairy can be avoided in small operations with proper care and procedures (see pages 196–197).

Strainer

A strainer is a necessity. It must be of a type that uses disposable milk strainer pads, available at any farm-supply store. (I won't even comment

Milking equipment includes a hooded stainless steel milk pail, strip cup, milk strainer with filter disks, milk can, and a hanging scale. Milk cans come in 2- and 4-quart sizes, and can be stainless steel or aluminum, but glass containers can also be used. Avoid plastic.

on the practice of running the milk through a hunk of cheesecloth, rinsing it out, and using it again.) The strainer must be small enough to fit into your holding container, such as a wide-mouth canning jar. Small tin kitchen strainers are inexpensive, and work well with small amounts of milk, but larger sizes made especially for backyard dairies are available in both stainless steel and aluminum. (Stainless steel costs about twice as much.)

If you're on a very tight budget, you can make a strainer from two large kitchen funnels. Cut off the spouts, and a little more, so you have two funnels with openings of 2 to 3 inches. Put a milk filter pad into one funnel, and place the other funnel on top to hold it down firmly.

Milk can be stored in 1-, 2-, or 4-quart glass jars, or in aluminum or stainless steel cans of the same size. Again, avoid plastic. Look for something easy to clean and sterilize.

Udder Washing Supplies
In addition to these tools, the home goat dairy will require a bucket to hold udder wash, or better, a spray bottle that will eliminate passing contamination from one goat to the next; an udder sponge or cloths; cloth or paper towels for drying the udder; and udder wash and cleansers and disinfectants for utensils. A scale for weighing milk so you can record production isn't an absolute necessity, but it's a very good idea.

Strip Cup
A *strip cup*, used to detect abnormal milk and mastitis, is simply a metal cup, with either a screen or a black tray at the top. You squirt the first stream of milk from each teat into the cup, and examine it for flakes, lumps, and other signs of abnormality. A strip cup will help you maintain a constant check on one aspect of your herd's health and the quality of the milk your family drinks.

Milking Stand
A milking stand is far more comfortable than squatting, especially if you have a number of milking does or if you tend to creak anyway. Milking stands can have stanchions to lock the doe's head in place to help control her while you're milking, and a rack to hold a feed pan to keep her occupied. A doe will quickly learn to jump up on the stand at milking time, especially if she knows there's grain waiting there.

How to Make a Folding Milking Stand

We've been using a folding, wall-mounted milking stand like this one for more than 40 years and highly recommend it. It's more comfortable to use than the common bench style (without the seat), and the folding feature makes it a real space saver. It can be built in a couple of hours for less than $40.

If, like most homesteaders, you have a "treasure pile" of recycled lumber on hand, the only expense would be the hardware. Unlike most milking stands, this one consists of two parts: a platform for the goat with a seat for the milker, and a stanchion to restrain the goat's head and to hold a feed pan.

Important Note: The nominal size of the stand is 42 inches long by 15 inches wide, but the size can be adjusted to fit your goats or the materials

MATERIALS & HARDWARE

MATERIALS

For platform

- 2 1 x 8 x 42" floorboards
- 2 1 x 4 x 15" cleats
- 1 1 x 8 x 32" seat
- 1 1 x 4 x 15" leg for platform
- 1 1 x 4 x 14" leg for seat

For stanchion

- 2 1 x 6 x 46" board (stanchion)
- 2 1 x 4 x 14" top cleats
- 2 1 x 4 x 14" bottom cleats
- 1 ¾ x 14 x 14" plywood for feed-pan holder

HARDWARE

- screws (or nails) (1¼" and 2¼")
- 2 4" T hinges for legs
- 2 4" strap hinges for stanchion
- 2 3" strap hinges for feed-pan holder
- 2 5" heavy strap hinges for platform
- 1 eye hook with 2 screw eyes for stanchion and stand
- 1 eye hook for feed-pan holder

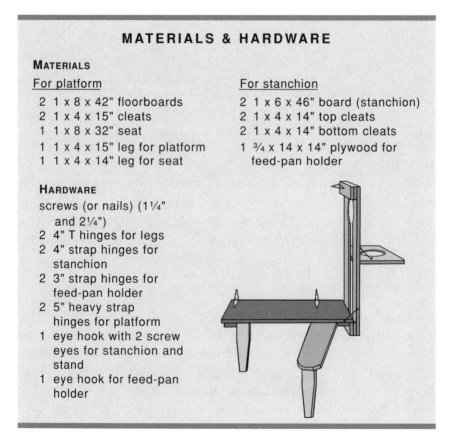

you might have on hand. Also, recall that modern lumber dimensions don't match the names: for example, a 1 x 6 board is actually ¾" x 5½". Such details aren't critical for this project.

Constructing the Platform

1. Lay the two 1 x 8 x 42" floorboards side-by-side (rough-side down, if you're using rough lumber that has one side better than the other).

2. Place a 1 x 4 x 15" cleat 1" from each end of the floorboards, as shown.

3. Fasten the cleats to the floorboards with screws or nails, making sure they don't extend through the floor. If nails protrude, bend them over and clinch them well.

4. Lay the 1 x 8 x 32" seat board on top of the platform so one corner protrudes beyond the plat-form by about an inch and the other side is flush with

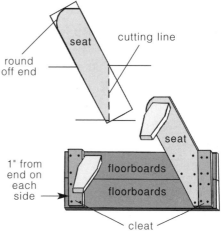

Steps 1–6

the cleat (see drawing). The angle of the seat isn't of extreme importance, but this method yields a good angle for most milkers.

5. Mark the seat with a cutting line along the cleat as shown. Cut the seat board so it fits snugly against the cleat. Trim off the other scrap.

6. Round off the end of the seat, and secure it to the platform as you did the cleats.

7. Taper the legs as shown, if you wish, or leave them square. Note that one leg is longer than the other.

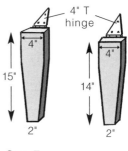

Step 7

8. Fasten the 1 x 4 x 15" leg to the underside of the platform, next to the rear cleat, with a 4" T hinge.

9. Fasten the 1 x 4 x 14" leg near the end of the seat with the other T hinge, as shown.

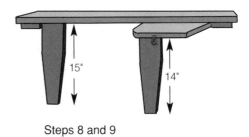

Steps 8 and 9

Constructing the Stanchion

1. Lay the two 1 x 6 x 46" boards about 4" apart, or so they line up with the width of the platform, minus 1". (If you have changed the platform dimensions, check this carefully, or the screw eye that holds the stanchion to the platform won't line up properly. The 1" difference allows the stanchion to fold over the platform.) For most goats, a neck space in the stanchion of 3½" to 4" is good.

2. Screw or nail two 1 x 4 x 14" cleats across the top and two across the bottom of the stanchion. (You need two cleats for each end so the stanchion can fold over the platform when it's not open for use.)

3. Center and draw an 8" circle just below the top cleat. Cut out as shown.

4. Center, draw, and cut out a circle in the 1¾ x 14 x 14" board that will hold the feed pan. The size of the feed pan you intend to use will determine the size of the circle.

5. Attach the two 4" strap hinges to the top and bottom cleats, as shown.

6. Using two 3" strap hinges, hinge the feed-pan holder to the other side, about 22" from the bottom of the stanchion. Attach hinges to the *underside* of the feed-pan holder.

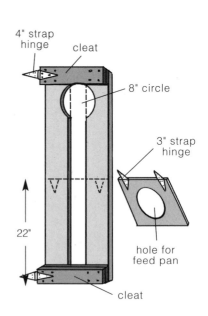

Mounting the Stand

1. You'll mount the platform to the wall first. Determine where the wall's studs are, and space the hinges accordingly. Attach the two 5" strap hinges to the platform.

2. Ensure that the platform is level before attaching it to the wall. Rather than measuring the distance from the floor to the hinges or from the floor to the platform, get the platform reasonably level *before* marking where the hinges will be attached to the wall, as the floor might not be level. If the floor slopes away from the wall, for example, the hinges might need to be less than 15" from the floor.

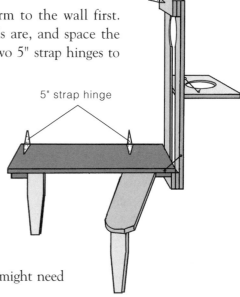

5" strap hinge

3. Secure the platform's hinges to the wall's studs.

4. Mount the stanchion to the wall. Ensure that the top of the stanchion's bottom cleat is snug with the underside of the platform for support. Also, be sure to leave a 1" space between the stanchion and the wall. (You need this clearance in order to close the stanchion over the platform when the unit is folded. This is why the stanchion is 1" narrower than the platform.) Screw the stanchion's 4" strap hinges to the wall, ideally through a stud.

5. Check that the platform and stanchion open and close properly and are reasonably plumb, square, and level.

6. Install an eye hook to hold the platform and stanchion together when the stand is open for business. Then fold the stand and, using the same hook, install another screw eye to hold the stand folded.

Mounting Tip

Check the stud spacing in the wall before mounting the milking platform to it. Screwing the hinges to studs provides a more secure installation. You don't want to attach the stand to drywall or plywood, for example.

7. Install an eye hook and screw eye to hold the feed-pan holder level.

Now you're ready to milk. Once your goats discover that there is feed in the pan, they'll readily jump up on the stand at milking time. When you're work is done, fold the stand out of ▶ the way.

Facilities: A Milking Parlor or In-Barn Milking

Ideally, the milking should be done in a special room, away from the goat pen and any sources of dust and odor. It should have good ventilation, running water, electricity, a drain, a minimum of shelves or other flat surfaces that gather dust, and impervious floor, walls, and ceiling so it can be kept even cleaner than your kitchen. All this, and more, is required for commercial grade A milk producers.

Alas, ideals are often unattainable for us poor goatherds. Many people with just a few goats milk in the aisle of the barn, *not* in the pen.

Dust, Bacteria, and Hair

Goat milk actually has less bacteria than cow milk as it leaves the udder. But on the debit side, the goat milk is more likely to pick up coliform bacteria during the milking process. This is due in part to the dry nature of goat dung. It actually becomes dusty. Combined with the loose housing most commonly used for goats, this results in dung dust and coliform bacteria in the air, on flat surfaces, and on the goat. And the goat milker is more likely to disturb the hair on the belly than the cow milker just because of the size of the animal. There might also be more of it, so hairy goats should be trimmed, even if it's only a "dairy clip" around the udder area during cold weather.

Preparing to Milk

Whether milked in a separate parlor or in the goat barn, the goat should be clipped, brushed to remove loose hair and dust, her udder should be washed with a dairy disinfectant and thoroughly dried, and a hooded milking pail should be used. It's also very important that the milker have clean, dry hands. Wash them before milking each doe.

Milking Procedure

Finally, we're ready for the actual milking. We'll take it step-by-step.

It looks easy, until you try it. But then with a little practice it *is* easy, and you'll wonder why you had milk up your sleeves and all over the wall and your legs the first time you tried.

Get into Position

Position yourself at the goat's side, facing the rear. Goats can be milked from either side, but they develop a definite preference for the side they're milked from. If you're right-handed, the right side will probably be easier. (You can forget about those pictures you've seen of people in foreign lands milking goats from the rear.)

Wash the Udder

The first step is washing the udder with warm water and an udder-washing solution, available at farm-supply stores. Follow the directions for the proper strength. Strong solutions can cause udder irritation. (Nowadays some people use alcohol-free baby wipes, which also help keep udder skin in good condition.)

Dry the udder and your hands with a paper towel, a fresh one for each goat, to avoid chapping and other udder problems, as well as to avoid milk contamination.

Draw the Milk

Grasp the teat with your thumb and index finger, encircling it near the base of the udder. Do not grasp the udder itself, which is sometimes tempting on goats without a clearly defined teat. That could cause udder tissue damage and the tissue can work its way into the teat with disastrous results.

Squeeze your thumb and index finger together to trap milk in the teat. This must be held firmly or when you squeeze the rest of the teat, the milk will be forced back up into the udder rather than into your pail.

Next, gently but firmly bring pressure on the teat with your second (middle) finger, forcing the milk down even farther. The third finger does the same, then the little finger, and if all has gone well the milk has no place to go but out of the teat — not necessarily where you want it, on your first try, but at least out of the teat.

The first squirt from each teat should be directed into a strip cup, a cup with a sieve or a black plate for a cover. That first stream is high in bacteria that have collected in the teat orifice and it shouldn't go into your pail. In addition, use of a strip cup will enable you to see any abnormality in the milk, such as lumps, clots, or stringiness. This is an indication of mastitis, which demands your attention (see chapter 8.) Never use the milk from any animal that's not in perfect health.

Then start milking into your pail, using first one hand on one teat, then the other hand on the other teat. With a little practice you'll develop rhythm.

Keep it up until you can't get any more milk, then massage or "bump" the udder, as kids do when sucking. You'll be able to get more milk. This massaging is important, not only because the last milk is highest in butterfat but also because if you don't get as much milk as possible the goat will stop producing as much as she is capable of.

The final step is stripping, or forcing out the last of the milk in the teat. (You never get all the milk in the udder.) This is done with the crotch of the thumb, or the tips of the thumb and forefinger. Grasp the teat at the top again, and force the milk out by running your fingers down the length of the teat. Be gentle; avoid vigorous stripping.

When you're finished, a teat dip is highly recommended. See the discussion under Mastitis in chapter 8.

Tip to Prevent Mastitis

Wash and dry your hands before and after milking each goat. Alternatively, use disposable latex gloves.

How to Milk in Eight Steps

1. *Close off the top of the teat with your thumb and forefinger so the milk flows out of the teat, not back into the udder.*

2. *Close your second finger, and the milk should start to squirt out. Discard the first stream. It will be high in bacteria.*

3. *Close the third finger. Use a steady pressure.*

4. *Close the little finger, and squeeze with the whole hand. Strive for a smooth, flowing motion. Don't pull on the teat. Just squeeze, gently.*

5. *Release the teat and let it fill up with milk. Repeat the process with the other hand.*

6. *When the milk flow has ceased, "bump" the udder, as kids often do while nursing, and you'll get a few more squirts. You never get all of the milk, but the last few drops can be extracted by stripping.*

7. *When stripping, take the teat between your thumb and first finger.*

8. *Squeeze gently down the length of the teat once or twice. Rough or prolonged stripping can injure the teat and udder, so some people skip this step altogether.*

Machine Milking

Goats can be milked by machine, but we won't discuss this in a book for beginners. While some people do use milking machines even for just a few animals, others claim they can milk dozens by hand in less time than it takes to set up and clean a machine. And that's without considering maintenance, the cost, or the noise. For most of us, milking is the most peaceful and therapeutic activity of the day. Why spoil it with an air compressor?

How Fast to Milk?

Some sources advise that the ideal milking speed for goats is 100 to 120 pulses per minute, whether milking by hand or by machine. This is faster than cows are milked, but best matches the rate at which kids suckle and results in better milk letdown, these people say.

Not so, according to veteran goat dairyman Harvey Considine, who has had as many as 1,000 goats at a time. He machine milks at 66 pulsations per minute.

Problem Milkers

There are problem milkers. If you've learned to milk with decent animals you can probably figure out how to cope with the other kind, but if you're unfortunate enough to have to learn on a troublemaker, some of the fun will go out of the experience.

First fresheners are most liable to be the culprits, although older does sometimes develop ornery habits, especially when they know you're using them to practice on. First fresheners are also likely to have small teats, which makes milking difficult, especially if you have large hands. On some, it's possible to milk by using the crotch of the thumb; others will require using the tips of the thumb and index finger in what amounts to stripping.

An occasional doe will have a tendency to kick, and almost any doe might kick once in a great while. This generally indicates that something is wrong. She's bothered by lice or flies, or you pinched her, or your fingernails are too long. Placing the bucket as far forward as possible, away from her hind legs, will help in this situation, and you can lean into her leg with your forearm to control movement. Leaning into the goat with your shoulder, holding her against the side of the milking bench or wall, will also serve to restrain ornery or nervous animals. It can also be useful in mild cases of "lying down on the job." Goats that have been nursing kids are especially prone to this sort of unhelpful behavior.

Milk Handling

Milk, especially raw milk, is highly perishable and extremely delicate. Following are some simple steps that will prevent spoilage.

- ◆ Cool milk immediately after milking.
- ◆ Don't add fresh warm milk to cold milk.
- ◆ Never expose milk to sunlight or fluorescent light.

Cooling milk immediately after milking means that it should not be left standing while you finish chores. Ideally, milk should be cooled down to 38°F within an hour after leaving the goat. That's quite a rapid drop when you consider that it was over 100°F when it left the udder. Home refrigerators aren't cold enough. Small containers may be cooled in the refrigerator, but anything more than a quart will not cool rapidly enough for good results unless it's immersed in ice water. (Remember, the bottom of the refrigerator is colder than the top.)

Don't add fresh warm milk to cold milk. If you're accumulating milk for cheesemaking, develop a system for rotating it, perhaps from left to right, or one shelf to another, so you know which is freshest.

If you store milk in glass jars, be sure to never leave them in the sun or fluorescent light, as this will change the flavor. But then, don't

To Cover or Not to Cover

Most "scientific" literature says it's absolutely necessary to keep the milk covered while it's cooling. However, a number of milkers have told *Countryside* magazine that their milk tastes better when it's left *uncovered* while cooling, or when the cover is left ajar. Cover it, and if your milk doesn't taste quite right, try it the other way.

leave a container of milk sitting out after a
meal in any event. Keep it cold. (See chapter 2
for more on off-flavors and other taste
problems.)

Cleaning Your Equipment

All equipment that comes into contact with
milk must be scrupulously clean. Cleaning
milking utensils is quite different from ordinary household dishwashing.
A dishcloth or sponge will not clean microscopic pores that hold bacteria
that will spoil milk or give it a bad flavor: a stiff brush must be used.
Household soaps and detergents contain perfumes that will leave a film
on equipment and may cause off-flavors in the milk. Household bleach
isn't pure enough for the dairy. You shouldn't even use tap water to rinse
milking equipment, or towels to dry it, because of the bacteria these contain. You need special compounds.

There are four dairy cleaning agents for dairy equipment. Two are for
washing: alkaline detergents and acid detergents. Iodine and chlorine
compounds are used for sanitizing.

The alkaline detergent is the basic cleaning agent. However, it leaves
a cloudy film called *milkstone*, which harbors bacteria. To get rid of the
milkstone you must use an acid detergent, which doesn't have the cleaning power of the alkaline detergent. Most dairy farmers scrub their equipment with alkaline detergent for 6 days, and on day 7, when everybody
else is resting, they scrub with acid detergent. If you have hard water,
which hastens the development of milkstone, you can put the acid detergent in the rinse water every day.

Chlorine compounds are used to sanitize equipment. You can also use
the iodine compounds that you wash goats' udders with (chlorine is far
too strong for the goats' skin) to sanitize equipment if you measure carefully and let the equipment soak long enough. At least 5 minutes is
required.

Measure all of the washing and sanitizing materials carefully. If the
solutions are too weak, they won't do the job they were intended for; if
too strong, you're wasting money and you run the risk of contaminating
your milk.

Taste Tip

For the best-tasting
milk, cool it to 35°F
within 30 minutes
after milking.

Here's the procedure for cleaning:

1. Rinse pails and other equipment as soon as possible with cold water to remove the milk proteins. Warm or hot water will "cook" the casein in the milk, leading to a buildup that can be difficult to remove. Never let milk dry in the pail.
2. Rinse with warm, but not hot, water and the alkaline, then the acid, detergent to remove the butterfat.
3. Wash in plenty of hot, soapy water, using a brush.
4. Rinse in plenty of hot water and chlorine or iodine sanitizing compound.
5. Invert on a rack, not on a shelf, to air-dry. Do not dry with a towel. Grade A dairies, by the way, are required to store milk buckets and containers upside down, without covers, on wire racks, to allow exposure to air. Why not do the same in your home dairy?
6. For extra insurance against bacterial contamination, just before using the equipment again, rinse it with a very small amount of chlorine bleach (diluted 10:1), then rinse with scalding water.

Commercial dairies must follow these procedures. The number of backyard goat raisers who go through all this is open to question. I've described them not because you'll croak from drinking milk that wasn't produced under hospital conditions, but so those who want to do a professional job will know they aren't doing it with soapy dishwater and a dishcloth and towel. If you ever encounter "bad" tasting milk, your milk handling and equipment cleaning procedures are the first things to examine.

Parting Shot

In grade A dairies, milk buckets and containers must be stored without covers, upside down on wire racks, to allow exposure to air. Shouldn't your home dairy do the same?

14

KEEPING RECORDS

Record keeping is necessary for the commercial goat dairy, because only through accurate and complete records does the owner know if the operation is making a profit — and if not, why not.

Record keeping is a necessity for the show goat breeder because only accurate and complete records will help to upgrade goats to the hoped-for blue ribbon status.

Most homesteaders and other backyard goat raisers shun records because they aren't involved with profit or upgrading or awards, and they think the work is a boring waste of time.

Big mistake. They're wrong on three counts.

It's true that the home dairy doesn't depend on goats for a living, as the commercial dairy does. But profits (and losses) show up in milk and dairy products that are better and cheaper than those purchased in the supermarkets. Even if the casual goat owner has no intention of ever entering a showring or even coming close to a goat show, it's still necessary to know certain facts about the herd's production and the results of management and breeding practices.

And record keeping can be fun! It becomes a challenge to have does that produce better than their mothers, and it's satisfying to look back on records that are several years old and see, in black and white, how you've progressed. No livestock breeder of any kind can afford to be without good records to use as a management tool.

If you have registered goats, pedigrees and registration certificates will be an important part of your files. The person you buy registered goats from will help you get started with these. There are several registries, with slightly different procedures. Get information on specifics from whichever one you choose to work with.

The Basic Barn Record

The basic barn record is a chart showing how much milk each goat produces. A plain sheet of paper with the goats' names written across the top and the days of the month down the left margin works fine. You can write the morning's milk in one corner of each square or imaginary square, and the evening's milk below it, as 4.5/4, with the first number the morning's yield of 4.5 pounds, and the second, the evening's.

Milk is measured by weight rather than volume, for official records on both goats and cows. It's the best procedure for the home dairy, too. Freshly drawn, unstrained milk foams, and it's difficult to gauge actual production in quarts, pints, or even cups. Then too, it's much simpler to deal in pounds and tenths of pounds rather than in quarts and fractions of quarts. For all practical purposes, a quart of milk weighs 2 pounds, and 8 pounds is a gallon.

It's a good idea to use this sheet to make notations of relevant data. For example, if you note "Susie in heat," you will be alerted to watch for her next cycle in 21 days. Notes on changes in feed, unusual conditions such as a doe not feeling well or acting quite right, or any other factor that might contribute to differences in milk production can be a big help in interpreting your records even years later. Even the weather can be important.

Any medications used — vaccines or wormers — should definitely be noted, including the type, amount, and date. And you might want to include hoof trimming and other routine chores on your calendar, too.

It's convenient to note breeding dates and the name of the buck, and freshening dates with all pertinent information, right on this same sheet.

Basic Barn Record

	CALPURNIA	CLEOPATRA	NEFERTITI	
March	MILK YIELD A.M./P.M.	MILK YIELD A.M./P.M.	MILK YIELD A.M./P.M.	TEMPERATURE/COMMENTS
1	3/3.25			32°, cloudy and windy
2	3/3			32°
3	3/3			31°
4	3/3.25			33° First geese
5	3/3			33°
6	3.25/3.25			32° First robin
7	3/3.25			38° (Vet) Autumn disbudded $24
8	3/3			54° Thunderstorm
9	3/3			28° Snow!
10	3.25/3	—		20° Cloudy 2 does! ☺
11	3.25/3.25	—		25° 100# feed $17
12	3/3	—		22°
13	3.25/3.25	2/1.5		21°
14	3.25/3.25	2/2		30° Gloomy
15	3.5/3.5	2/2		35° ½" rain — ice off pond
16	3.5/3.5	1.5/2		11° Clear — ice on pond again
17	3.5/3.5	1.5/2		21° Overnight low 8°
18	3/3.25	2/2		32° 1" snow 6 bales hay $12
19	3.25/3.25	2/2		34°
20	3.25/3.25	2.25/2.25		35° Dreary 13 chicks hatched
21	3.25/3.25	2.5/2.25		39° Foggy
22	3.25/3.25	2.5/2.5		45° Still gloomy
23	3.25/3.25	2.5/2.75		50° Light rain (at last!)
24	3.25/3.25	2.5/2.5	—	48° 2 bucks ☹
25	3.25/3.25	2.75/2.5	—	40° 100# feed $17
26	3.25/3.25	2.5/2.75	—	40°
27	3/3.25	2.5/2.5	1/1.5	35°
28	3/3	2.75/3	2/2	32° Overcast, flurries
29	3/3.25	2.75/2.75	2/2	35° Cloudy
30	3/3.25	3/3	2.25/2.5	35°
31	3.5/3.25	3/3	2.25/2.5	35°

This is an example of a very simple barn record form. Real barn records are usually hastily scribbled, often stained, wrinkled, and I've even had a few with goat teeth marks on them. The basic information is milk production, which is weighed and recorded at each milking. Other data such as feed purchases, medications administered, kidding dates, and even weather conditions can be included. To make these records even more valuable, they can later be transferred to a spreadsheet so you can determine the cost of your milk, create individual production records, and even chart lactation curves for each of your goats.

When we bought feed instead of growing our own, that went on the sheet, too. At the end of the year we had a complete record of the input, output, and interesting happenings in our barn, all on just 12 pages.

Recording Reality

One of the primary advantages of such a system is that it overcomes the natural forgetfulness of most human brains. Let's face it: few people, if any, are going to remember the statistics from 730 trips to the barn a year that, if the herd consists of three goats with lactations of 305 days, means 1,830 separate entries each year for milk alone.

In addition to not being able to remember all those numbers, the brain can distort them. For example, you might be impressed by Susie's production of 1 gallon of milk in 1 day and consider her the best goat in your herd. But your records might indicate that a less spectacular producer that just chugged along less dramatically, but with a long and steady lactation, actually produced more than the flashing star. If you had to cull milkers or make a decision about whose daughter to keep you might make the wrong choice without those records.

In many cases, breeders will note that a relatively few top does produce as much as a larger number of poorer producers. Since poor producers require just as much work as good ones and eat virtually as much, it follows that milk from the lower third or even half of your herd costs more than milk from the top half or two-thirds. This suggests that you could get more milk for the same amount of time, effort, and money by replacing the poor does with daughters from the best does. If you don't need that much milk you might be able to eliminate one or more animals, reduce your feed bill, and still increase your milk production.

Records of breeding, expenses, income, and milk production are all basic, and it doesn't take much time or knowledge of accounting to keep them. Just the act of writing them down will tell you a few things about your operation, but it's also possible to squeeze a lot more helpful and interesting information out of those records. Today, with computers, it's easier than ever. Check the publications for programs written specifically for goats, or simply use a spreadsheet.

In any given category, your cost and income will vary according to your location, your type of operation, and your management methods. Even when these factors remain constant, your costs can vary from

The Basic Balance Sheet

Of course, you don't need a computer to tell you how much your goats are worth to you. You can keep monthly and annual tallies of a few simple income and expense items. Here's an example

INCOME

Milk sales	$_____
Sales of stock	$_____
Family milk	$_____
Family meat	$_____
Stud fees	$_____
Boarding fees	$_____
Misc. (disbudding services, etc.)	$_____
TOTAL	$_____

EXPENSES

Purchase of stock	$_____
Feed (grain, minerals, salt)	$_____
Hay	$_____
Veterinarian and medicine	$_____
Repairs on equipment	$_____
Supplies	$_____
Advertising	$_____
Registry, transfers, etc.	$_____
Telephone, postage	$_____
Amortized costs	$_____
TOTAL	$_____
PROFIT (or loss) (income – expense)	_____

Note: This record can be much more detailed. You might want to divide "Sales of stock" into breeding stock, meat animals, and perhaps pets, to learn where your best market is. It could be very helpful to separate hay from grain. For example, you might look at your records, determine that you're spending too much on hay, and decide to build a new, hay-saving manger. Later, you'll want to use those records to see how much money the new manger is saving you. And if it works really well so the goats are no longer bedding themselves with wasted hay, you might want to add a category for straw!

year-to-year. In a drought year, the price of hay can double. You might have mostly doe kids one year, with heavy registration expenses. The next year you might have mostly bucks that go to the butcher, and very low registration expenses. Medical expenses often come in spurts. And don't forget that if you produce more milk than your household can use and the surplus is used to feed calves or pigs, there should be a price differential.

Figuring Out Costs

It would be helpful to know in advance how much a gallon of milk from your home dairy will cost, perhaps as a way to justify the investment in a goat in the first place. Unfortunately, there is no reasonable answer to that. Even well-managed commercial dairies have costs that vary widely. Dairy plants that buy goat milk likewise vary in their payment schedules from place-to-place, and even month-to-month, as well as on protein and other factors. Recently, this has ranged from not much more than $20 per hundred pounds to as high as $43. This is from $1.60 to $3.44 a gallon! The cost to produce this milk varies just as widely in commercial herds, and perhaps even more in home dairies.

That's precisely why you should keep records. Your own balance sheet is the only one that counts.

There's no point in looking at numbers and setting unrealistic goals for yourself — or being disappointed because you think you don't measure up. You should determine what your milk is costing you and balance that against what you're saving at the grocery store. But don't forget to add in the value of meat, fertilizer, the security and pleasure of providing your own dairy products, and the fun of having goats!

Pricing Milk

Pricing the milk from your home dairy is no simple, cut-and-dried calculation. Even if you go by the maxim that anything is worth only what someone else is willing to pay for it, goat milk presents special problems.

Most of us produce milk that has varying value. In winter, when production is likely to be low, the entire output might be used for drinking (the "fluid milk" market, the dairy industry calls it). If we'd be willing to pay the health food store price of fresh goat milk, this milk is quite valuable. If you have a baby who's allergic to cow milk and can't find goat milk — at any price — the cost of your home-produced milk is probably of little concern. If, without goats, we'd be drinking cow milk, it's somewhat less valuable. As the milk flow increases we might begin to use some to make cheese, or yogurt. This is "manufacturing" milk, and even cow farmers get less for it. When we begin to use an even greater surplus as feed for pigs, calves, or puppies, the value slides even further.

That's not all. Do you consider manure to be "waste," or is it black gold? Do culls go to the butcher, or to the rendering plant? Do you utilize the hides? In one way or another, all of these affect the cost of your milk.

In other words, you have a lot of leeway in putting a price on the milk you use yourself. Remember, you're not keeping these records or coming up with numbers for the bank or the IRS. They're strictly for your own use, valuable tools that will improve your herd's performance.

Capital Costs and Operating Expenses

Figure your *capital* costs, that is, money you invested in things that aren't used up all at once. This includes milking equipment, feed pans and water buckets, fencing, tools such as the disbudding iron and clippers and tattoo set, and the goat itself. Naturally you don't want to charge all this against the milk produced in 1 year.

The milk pail might last 20 years: take one-twentieth of the price as this year's cost. (That's conservative: ours is going on 40. I have long forgotten what we paid for it.) The goat might be good for another 5 years: take one-fifth of what you paid for her. Go down the list of capital goods, determine the capital costs for 1 year, and you'll have a more honest picture of your true costs.

Then add up your operating expenses: hay and grain, electricity used in the barn, veterinary fees, milk filters, and everything else that was purchased and used up.

Add up the operating expenses and 1-year cost of capital equipment and stock, and subtract that from the value of the goat's production. This will give you a pretty good idea of the goat's annual value to you. By adding up all these costs and dividing by the number of quarts or gallons of milk produced, you'll know the actual cost of your milk.

Even this isn't completely accurate, but it's adequate for most people, and far better than a complete disregard for accounting. If you're inclined, you can figure in the cost (or value) of labor, the value of manure, the cost of money or the return on investment, taxes, and even more.

If more goat raisers kept such records, you can be sure there wouldn't be very many $10 or $20 goats for sale! More people would pay better attention to culling and proper management, too, if they knew what their goats were really costing them.

The Market for Goats

Some sources suggest that many herds break even not because of the value of the milk but because of the value of the kids. A purebred and registered herd that can command top price for its animals will come out far ahead of a herd of grades whose kids are a drug on the market.

This is still true, but notice the wording: "*can* command top price." At a particular time in a particular place, even excellent purebred stock doesn't always bring high prices. This could be simply because there are plenty of good grades available, and that's what most people in the area want anyway. Still, many other factors could be involved. For top prices you need not only top animals, but you have to earn a reputation in the showring, you'll probably have to be on test and have your animals classified, and you have to advertise. All of these require time, money, energy, and often frustration. How can you know if it will pay off?

One way is to make projections, based on current prices and costs and market conditions in your particular area. This is just one more way of making records work for you. On the other hand, many goat owners simply aren't interested in increasing profits if it means going to shows, getting involved in registering animals and all that entails, or dealing with the public. This can be reflected in their bookkeeping by emphasizing what's important to *them.*

CHEVON

In Sir Walter Scott's *Ivanhoe*, Wamba the jester gives Gurth the swine-herd an English lesson. While domestic animals are alive and must be tended, he explains, we use their simple Saxon names: calf, cow, sheep, pig, and so on. But when they are dressed and served to the Norman conquerors, they are called by their Norman (French) names: veal, beef, mutton, and pork.

In twentieth-century America, the equivalent of the Norman conqueror was the consumer. People weren't very interested in eating *goat*, so goat farmers made up a new word that would be more appealing: *chevon*, from the French words for goat, *chévre*, and sheep, *mouton*. (Goat cheese is also called chèvre.) Of course, we might also call goat meat *cabrito*, from the Spanish for "little goat," but except in the Southwest, *chevon* is the more common term.

The Market for Chevon

Goat meat by any name is very popular in many cultures. In the United States the popularity of goat meat used to be confined mainly to people of Spanish, Greek, and Jewish heritage, but recent Asian immigrants have greatly expanded the market. Kid is an important part of the meals for spring festivals of several religions, but chevon is a traditional — and delicious — everyday meat for many.

While most meat goats are raised and slaughtered in the Southwest, where they can be raised cheaply on range and where a large market exists, the market in the Northeast has grown rapidly and continues to expand. In 1997, an estimated 30,000 goats a month were shipped into New York City. More than 60 percent of the goat meat sold in New York is purchased by Muslims, but there are plenty of other kinds of good customers. Most of these animals originate in Texas, Tennessee, Missouri, and Oklahoma. As of 1999, the University of Kentucky Extension was encouraging farmers in that state to produce meat for the New York market.

But almost every region has at least a small niche market for goat meat today. This is one reason for the soaring popularity of meat goats such as the Boer: Boers and Boer crosses grow much faster and are "meatier" than dairy breeds.

What Does Chevon Have to Do with Dairy Goats?

Meat is an important by-product of dairying. Over the years, any farm will average 50 percent buck kids. Not one in a hundred can be kept, profitably, as a herd sire. While there is a limited demand for wethers (castrated males) as pets in some areas, it is more merciful in most cases to butcher them for meat.

In addition to unwanted males, any dairy operation will have cull or aged does that simply are not paying their way. Resist any temptation to sell them as milkers to someone else. You might make a few dollars on the deal, but the cost to the goat world — and to your reputation — will be far higher.

Culling is a fact of life when dealing with livestock, but that doesn't make it any easier — especially for city people with no livestock experience, and even more so when dealing with animals like goats! Butchering is never a pleasant task, and it's normal to have qualms about eating an animal you raised yourself. However, once you overcome any initial reticence you'll most likely agree that chevon is a delicious bonus of your home dairy.

Slaughtering and Butchering

Goats are commonly slaughtered at one of four stages:

- On farms where milk is valuable or where the labor required to raise kids is deemed out of proportion to the value of the meat, kids may be butchered at birth and dressed like rabbits.
- Milk-fed kids weighing between 20 and 30 pounds are the most popular for the Easter-Passover market.
- Castrated buck kids may be kept and butchered at 6 to 8 months for a goat-meat market or your home table.
- Cull does can be processed into jerky, salami, or anything that makes use of meat that isn't especially tender.

Slaughtering Tip

Withhold feed, but not water, for 24 hours before slaughtering. Butchering will be easier if there is no feed in the paunch.

Dispatching the Animal

If you dispatch the animal with a gun, aim from the back so as not to frighten it by the sight of the barrel. Some people prefer to use a hammer. A sharp blow to the skull will render the animal unconscious, and the jugular can be cut with a sharp, stout knife.

Hanging and Bleeding

Although some people claim it's merely a cultural practice, most will say that thorough bleeding is important. Do not damage the heart in the killing process so it can continue to pump blood, and hang the animal head down to allow complete drainage. If you do much butchering, a gambrel hook will prove useful, but a carcass can be hung by passing a metal or wooden rod through the tendons of the rear legs, or even by tying it to a rafter or a tree branch with a rope, as many deer hunters do.

If you are, or know, a big game hunter, you might have access to a gambrel hook like this one. But if you're only going to butcher one or two goats each year you can make do with a piece of wood such as a tree branch. Or simply suspend the carcass with rope.

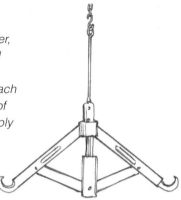

Skinning

We've been told that the Greeks, who have a long tradition in goat butchering, cut a small incision between the hind legs and blow up the hide like a balloon. This helps separate the hide from the meat and makes skinning easier and cleaner. No doubt most people would rather use a tire pump than their mouths, but one goat raiser reported inserting the nozzle of a garden hose into the incision and filling the space with cold water. In addition to separating the hide, this helped to cool the meat.

To skin the animal with or without this step, carefully cut a slit from between the hind legs to the throat. Don't cut too deeply, as you don't want to cut into the intestines or meat. Once started, you can usually work your fingers beneath the skin to hold it away from the body.

From the two ends of this cut continue out along the insides of all four legs. The skin is tighter on the legs, and again, try not to cut into the meat.

The pelt will be attached at the tail. To remove that, cut around the anus and loosen it until you can pull out a length of colon. Tie off the colon with a piece of strong string to avoid possible contamination. Cut it off above the string and let it fall back into the body cavity. Then cut off the skin at the base of the tail. If you're very frugal, skin out the tail and use it for stew meat.

If you're saving the hide, cut it off as close to the ears as possible. Skins from newborn goats are more like fur than hide, and many useful items can be made from any tanned goat skin. If you aren't interested in keeping the skin, cut the head off with the skin. In either case, remove the head at the base of the skull.

Butchering

When the hide is removed, cut down the belly, from between the hind legs to the brisket. If the animal was starved for 24 hours before slaughter (again, don't withhold water), the paunch will be empty and there will be less chance of cutting into it, but be careful anyway. Let the paunch and intestines roll out and hang.

Work the loosened colon end down past the kidneys, and carefully remove the bladder. Pull out the liver and remove the gallbladder by cutting off a piece of the liver with it. If the gallbladder breaks and spills bile on the liver, wash the meat in cold water immediately to avoid a bitter taste.

The offal will fall free when the gullet is cut.

Saw the brisket — an ordinary carpenter's saw will work if you don't have a meat saw — and remove the heart and lungs. Clean out any remaining pieces of tissue, wash the carcass with cold water, and wipe it dry.

The skull can be split to get the brains, and the tongue removed. Wash the liver, heart, and tongue in cold water and drain them.

Cutting, Dividing, and Packaging the Meat

Newborn goats weighing about 8 pounds can be cut up like rabbits. Cut through the back just in front of the hind legs and again just behind the front legs. Each of these can be cut along the spine, giving you six pieces of meat.

Larger animals, cut like lamb, will yield roasts, chops, ribs, and trimmings that can be ground and used in patties or mixed with pork for sausage. Some of the larger pieces such as the legs can be cured like hams, or corned.

If you have little or no butchering or meat-cutting experience, the thought of converting a carcass into neat packages like you see at the meat counter might seem like a formidable task. But don't let it throw you. Simply cut off pieces that "look about right to be a roast" or wherever, and they'll be just as edible and probably even better-tasting than those from the supermarket. You really can't do anything "wrong" at this point.

Ideas for Cooking Chevon

If you enjoy lamb, use your favorite lamb recipes with chevon. Since we raise both sheep and goats we often have both in the freezer, and it's hard to tell the difference.

There are many ethnic dishes from such areas as Greece and Turkey that call for chevon. Oregano is a good spice to use with chevon, and the meat is excellent in curries. You'll find a few suggestions for getting started with chevon cookery in chapter 17.

DAIRY PRODUCTS

Great-tasting fresh milk is the primary reason for owning a dairy goat. But since goats are like potato chips (you can't have just one), and because of the fluctuations in milk production, sooner or later you're going to have more milk than your family can possibly drink. Then it's time to think about other dairy products.

When I started with goats, yogurt was unheard of in our neck of the woods. Today it's practically a staple, and many people make their own from store-bought cow milk. And in the past few years, goat cheeses have become immensely popular. So actually, today, you might have thought about those even before getting your first goat. Even then, you might not yet be aware of all the possibilities.

One of the joys of having goats is the dairy products you can make in your own kitchen. At certain times of the year, you will have a surplus of milk that can be turned into a variety of products that will make you more independent of the supermarket and will make your goats more valuable to you. It doesn't make sense to produce wonderful milk and then throw half of it out because you can't drink it all. And perhaps most important of all, making cheese and other dairy products is satisfying and a lot of fun!

Preserving Milk for Future Needs

When your goats are dry, it certainly isn't fun to have to buy hay and grain *and* milk. If you know you're going to run out of milk later, you'll probably want to consider preserving some fluid milk.

Frozen Milk

Freezing milk is simple. This can also be a good idea if you don't have enough to bother making cheese, and already have enough yogurt.

Freeze it in plastic jugs, leaving an air space for expansion. Thawed frozen milk is somewhat watery, and while it's fine for cooking, you'll probably want to mix it with fresh milk for drinking.

Canned Milk

Although it's controversial, milk can also be pressure canned. Home economists say that canning milk at home is dangerous. Customer service representatives at several companies that manufacture pressure-canning equipment have told me they had no information on canning milk, and that you can't can milk because "it will curdle." Some people who have been canning milk for years say this is all nonsense. You'll have to make up your own mind, but should you decide to try it, here's how:

1. Follow your regular pressure-canner instructions regarding the amount of water to use, allowing steam to escape before closing the vent, and so on.
2. Fill canning jars to within 1 inch of the top with fresh, warm (120°F) milk.
3. Add sterilized lids and rings.
4. Process for 12 minutes at 12½ pounds pressure. (Another method calls for 25 minutes at 10 pounds of pressure.)
5. Let the jars cool in the canner, undisturbed.
6. Remove cans, and store in a dark place.

Canned milk has a slight caramel-like flavor, so it's best used for cooking. Also, the butterfat comes to the top and the calcium settles to the bottom in canned milk. Just shake the jar before you open it.

Lacking a pressure canner, process the jars in a hot water bath for 3 hours. (This method is definitely not recommended by home economists, but the milk might be used to feed kids or other animals.)

Evaporated and Condensed Milk

If milk is heated to about 190°F and then simmered very slowly, the water will evaporate. When it's reduced by half, it's evaporated milk. Simple. The kicker is that this can take up to 2 days. If you want to try it, probably on a woodstove and when you can use the humidity, use a large double boiler or one of those insulating pads that provides an air space between the heat and the pot to prevent scorching. For larger quantities, improvise a double boiler by putting about 2 cups of milk in each quart jar, place the jars in a canning kettle containing about 2 inches of warm water, and add enough water to reach the level of the milk.

Condensed milk can be made the same way by mixing 2 cups of sugar into each 2 cups of milk. One cookbook author advises only 1 part sugar to 2 parts milk and still calls it "excessively sweet." She also says that if the milk is simmered, stirring often, it will be ready in about 2 hours.

Dried Milk

Most people say you can't dry milk at home. After investigating the commercial process, I agree. Even Mary Bell, who wrote *Mary Bell's Complete Dehydrator Cookbook* and dries such far-out things as pickles and watermelon, doesn't dry milk.

However, Mary Jane Toth's *Caprine Cooking* passes on a method furnished by one Jill G. Simkins, who, we assume, has done it successfully. Here's the method: She simmers 1 to 2 gallons of milk in a double boiler until it's evaporated to the consistency of cream. She pours that into a large pan and dries it in the oven, with the door ajar, at about 150°F. The dried product is then ground in a blender. To use it, she soaks 1 part milk in 4 parts water.

Cheese Making

Sooner or later you'll want to try making cheese. In its most basic forms, making cheese doesn't require much equipment, it takes little actual working time, and it can become quite a hobby, or even a business. There is as much art and science to making cheese, however, as there is to making wine. Don't expect to come up with any "vintage" cheeses without experience or some luck. Neither of these helps me very much, but it's fun anyway, and the cheese is at least edible — usually!

But remember that the very first cheese was probably a "mistake." The theory is that some ancient unsung shepherd carried milk for his lunch in a pouch made from a kid's stomach, which is where cheese-making rennet comes from. The milk formed a curd, which the shepherd was either curious, hungry, or dumb enough to taste. The rest is history.

Over that history have evolved many hundreds of supreme cheeses, virtually all developed and perpetuated by rural homesteads, and probably as many by accident as by design. Modern technology only tries to imitate those homestead cheeses.

Until recently, however, we homestead-types usually confined our cheeses to a very few, very simple types, mostly soft, fresh cheeses. Aging or curing is an additional step that requires more interest and dedication, and a little more equipment. Today, thanks to a number of people who have taken home cheese making to new heights, almost anyone can make dozens of specialty gourmet cheeses at home. A number of mail-order companies now supply everything you could possibly need in the cheese department, and lots more. Yes, you can make cheddar, mozzarella, feta, and brick, but also several forms of chèvre, gouda, *queso blanco,* and hundreds of others, including many that gourmets rave about.

The Basics of Cheese

In its simplest forms, cheese is nothing but curdled milk with the whey drained off. But starting with this, well over five hundred named cheeses can be made.

The milk can be curdled with animal rennet (an enzyme), vegetable rennet, acids such as vinegar or lemon juice, or with a combination, each producing different results. Curdling requires heating, but a few degrees' difference in temperature can affect the end result, as can the length of the heating period. Starter cultures can be added, the specific bacteria in each culture contributing still more differences. Some cheeses are pressed; others are not. Those that are can be held for variable times, under variable pressures. The temperature and humidity in the kitchen, the size and shape of the cheese, if and how it is aged — all this and more determines what kind of cheese that basic curdled milk will become.

Oh, and don't forget the milk! The composition of the milk can be a factor in how your cheese turns out, but the most important thing to remember is that you can't turn bad milk into good cheese. Always use fresh, good-tasting milk.

As you can see, cheese making can be much more complicated than goat raising. If you really get into it you'll want to study books and articles devoted to it, search the Web, perhaps take classes, and network with other cheese makers. You will find several and, sometimes, many recipes for cheeses with the same name but slightly different procedures, and it's likely that each will be different. Only experience will tell you which ones are best for you. But here are a few simple recipes to get you started.

To Pasteurize or Not?

You should be aware that the use of pasteurized milk in cheese making is another hotly debated topic, although with slightly different nuances than the raw milk for drinking controversy. Many modern recipes call for pasteurized milk, which means that the naturally occurring bacteria, good as well as bad, have been killed. Then the cheese maker adds specially chosen bacteria to produce a certain kind of cheese. This provides more control over the process, but raw milk can produce more exquisite cheeses that even peasant (as well as snobbish) gourmets rave about.

It's interesting to note that after 2 months of aging, cheese develops antibiotic properties that kill almost all disease germs that might have been in the milk. I recall reading a news item several years ago about a cheese, probably goat, found at an archaeological dig, that was said to be several thousand years old, and was still edible. Don't ask me how they knew it was edible!

Basic Cheese

You can make several types of basic cheese with equipment you probably already have in your kitchen, plus a dairy thermometer. A large stockpot, stirring spoon, colander, and cheesecloth along with lemon juice or vinegar and, of course, some milk, will get you started.

Vinegar Cheese

Milk that is several days old often works best with this recipe.

2 quarts milk
¼ cup vinegar, lemon juice, or lime juice

1. Heat the milk to 185°F, or close to boiling.

2. Add the vinegar very slowly, while stirring. The milk will start to coagulate as soon as you add the vinegar, separating into curds and whey.

3. When curds appear, skim them off into a cheesecloth-lined colander. Or pour the curds and whey into the colander. (This is easier, but skimming results in a better flavor and texture.)

4. Tie the corners of the cheesecloth together, and hang the cheese where it can drain for several hours, or until it stops dripping.

5. Slice or cube, and eat it as is, although it's rather flavorless. Or make into *queso blanco*. (See recipe on page 218.)

Ricotta

Make ricotta by following the recipe for vinegar cheese above, but use whey instead of milk. The whey, of course, is left over from making other cheese: you get two for the price of one!

Queso Blanco

*Bland vinegar cheese becomes Mexican queso blanco (white cheese).
Since it doesn't melt when cooked, you can use it
in stir-fry dishes in place of tofu, or fry 1-inch cubes in hot oil,
and serve with a yogurt dip for a special snack.*

Vinegar cheese
Your choice of herbed olive oil, spiced wine, or soy sauce
or chopped pimentos, green chilies, jalapeños, pitted
ripe olives, and salt

1. Make vinegar cheese (see page 217).

2. Marinate cheese in olive oil, wine, soy sauce, or any concoction
that pleases your palate or imagination. Or mix in pimentos, green
chilies, jalapeños, olives, and salt to taste.

3. For a product firm enough to slice, press cheese for 8 to 10 hours.

Cottage Cheese

*To make cottage cheese, you replace the acid with rennet, change the heating
procedure a little, and treat the curds a bit differently.
The resulting cheese is delicious as is, or add salt, cream, chives,
or whatever you like. It should keep about 1 week in the refrigerator,
but once set on the table, ours never lasts nearly that long.*

1 gallon goat milk
¼ tablet rennet
½ cup cold water

1. Warm the milk to 86°F.

2. Crush the piece of rennet tablet with a spoon, and dissolve it in the
cold water (or follow the directions that come with the rennet you
use, liquid or tablet).

3. Add the rennet solution to the milk, stir slightly, and let stand in a warm place until a curd forms, usually about 1 hour.

4. Test to see that the curd "breaks" by inserting the dairy thermometer or other object 1 inch or so, lifting it out, and making sure the curd has clean edges.

5. When the curd breaks, cut it into small cubes with a long, thin-bladed knife (*a*). (A serrated bread knife works very well.) Make a graphlike pattern by cutting the curd vertically (all the way through to the bottom of the pot) into ½- to 1-inch squares (*b*). Then slant the knife about 45 degrees, and make slices at right angles to the first ones (*c*), reaching to the bottom and sides of your pot. (See illustrations below.)

6. Stir the mass, very gently, and cut any large pieces remaining.

7. Warm the curds and whey to 110°F, very slowly and gently. Within limits, the longer it takes to get to this temperature, the firmer the curd will be.

8. Pour the curds and whey into a colander lined with cheesecloth, and drain for a few minutes.

9. Run cold water over curds and whey to rinse off the whey. This gives it a milder flavor and increases its shelf life.

A Harder Cheese

This is the first cheese we made, following the directions that came
with the rennet. We have made it in countless variations since then.
It takes about 10 pounds of milk to make 1 pound of cheese,
so I don't like to mess around with less than 6 quarts (1 ½ gallons)
of milk, but that's up to you. One-quarter of a rennet tablet will work
for most any quantity of milk you'll use in the kitchen.

1½ gallons goat milk
¼ tablet rennet
½ cup cold water
Salt

1. As for cottage cheese, warm the milk to 86°F, crush the rennet
   tablet, dissolve it in the water (or follow rennet instructions), add
   the rennet solution to the milk, and stir slightly.

2. To curdle the milk, keep it at about 90°F: place your cheese kettle
   inside a larger container of water at about that temperature.

3. When the curd is set, cut it as described in the cottage cheese recipe
   on page 219, keeping the cubes on the small side, ½-inch square or
   smaller. Usually, the smaller the curd, the harder the cheese.

4. Warm the mass to 130°F very slowly, stirring occasionally to keep
   the curds separated. This should take 1½ to 2½ hours for the curds
   to reach the right consistency. If the curds are not firm enough
   when you remove them, the cheese will be pasty and sour. If they're
   too firm, the cheese will be dry and crumbly.

5. Pour the curds and whey into a large container lined with cheesecloth. Lift the cheesecloth, bring the four corners together, tie
   them, and let the curds drain into a container. (We tie a string to
   the corners and tie that to a cupboard-door handle. The cheese
   kettle fits nicely on the countertop, beneath the bag.) Draining
   should take about half an hour.

6. If you have a cheese press, use it. If you don't, you can easily devise
   a homemade press by following steps 7 through 10.

7. Place the mass of curd on four thicknesses of cheesecloth. Wrap a band of cheesecloth, about 3 inches wide, very tightly around the width of the cheese, and pin the cloth to itself with a safety pin.

8. Press down on the cheese to fill out the cloth band and form a small, flat wheel. Gently rub the top to give it a smooth finish.

9. Cover with a damp cloth and a board or plate. Weight the board or plate with a scrubbed brick, stone, or whatever you have.

10. Turn the cheese over before you go to bed and replace the weight.

11. The next morning, remove the cheese from the cloth, sprinkle the cheese all over with salt, and rub it in. Place the cheese in a clean, cool place. The ideal temperature is 56°F.

12. The next day, turn the cheese again and rub in the salt.

13. Turn every 24 hours for about 2 weeks, then every other day for another 2 weeks. It's ready to eat after these 4 weeks, but if everything went right and it didn't develop cracks or molds, it will improve with age.

Cheddar

One form of cheddar can be made by following the basic recipe for harder cheese (see page 220), up to pouring off the whey.

1. Follow the recipe for A Harder Cheese through step 5.

2. Place the cubed curds in a colander. In an oven or a double boiler, heat to 100°F for 1½ hours.

3. When the curd has formed a solid mass instead of the individual cubes you started with, slice it into 1-inch-thick strips. Dry for 1 hour, turning every 15 minutes to allow even drying and holding the temperature at 100°F.

4. Salt the curd as in step 11 of the basic harder cheese recipe, and follow the rest of the steps in that recipe. Curing is much longer, however; cheddar takes 6 months to cure.

Feta

One of my favorite variations is a form of feta.

1. Follow the recipe for A Harder Cheese recipe (page 220) through step 10.

2. After the cheese has been pressed and the weights and the cheese-cloth removed, instead of curing it, cut it into cubes about 2 inches square.

3. Soak the cubes in jars of superconcentrated brine (more salt than will dissolve in the water) for a day.

4. Store in a 14 percent brine solution.

Gouda

An easy hard cheese to start with, suggest Bob and Ricki Carroll of New England Cheesemaking Supply, is Gouda, named for the region in Holland where it originated. Gouda can be enjoyed after only 2 months of aging, but it develops a superior flavor after 6 months or longer.

2 gallons goat milk
6 ounces buttermilk culture, made 1 day in advance by sterilizing 1 quart goat milk for 30 minutes in a water-bath canner, cooling to room temperature, adding a packet of Hansen's buttermilk culture, and setting for 12–24 hours
Few drops liquid cheese coloring
½ teaspoon liquid calf rennet
½ cup cool water

Cheese Waxing Tips

♦ If the cheese is too moist, the wax won't stick.

♦ If the wax forms a thick layer, it's probably too cool.

♦ If you don't get a definite layer of wax, then the wax is too hot.

1. Heat the milk to 86°F.

2. Add the buttermilk culture. Keep the milk at 86°F for 30 minutes.

3. Add the coloring, unless you don't mind a Gouda that looks somewhat like lard.

4. Add the rennet to the cool water, and gently stir this solution into the 86°F milk. Continue to maintain this temperature for 30 to 45 minutes. (Energy saving tip: Put the cheese kettle in a larger pot or sink full of water that's about 86°F.)

5. Check that the curd is firm and has a "clean break": insert the dairy thermometer into the curd at an angle and see if the curd breaks cleanly. If it does, it's ready to be cut.

6. Cut the curd into ¼-inch pieces, as described on page 218 in the cottage cheese recipe.

7. Let the curd set for a few minutes, until it sinks into the whey.

8. Ladle or carefully pour whey out of the kettle until the curds are visible.

9. This is a step peculiar to Gouda: add a quantity of water heated to 175°F, until the curd-whey-water temperature reaches 92°F. The amount of water isn't important; it's the temperature of the curd that counts. Let sit for 15 minutes, stirring often. The curds will become smaller and harder.

10. Let the curds settle to the bottom, drain off the whey until the curds are visible, and add more water at 175°F. This time bring the temperature of the curds and whey up to 96°F, and maintain that temperature for 30 minutes.

11. Drain the curds, and quickly place them in a cheesecloth-lined mold. Press with 30 pounds of pressure for 15 minutes.

12. Remove the cheese from the mold, turn it over, and put it back into the cheesecloth-lined mold. Press for 6 hours with 50 pounds of pressure.

13. Remove the cheese from the press, remove the cheesecloth, and let the cheese set in the mold overnight without pressure.

14. The next day, make a brine by adding salt to a gallon of water until no more salt will dissolve. Soak the cheese in the brine for 2 to 3 hours.

15. Air-dry the cheese until the rind is dry to the touch. This can take 1 to 3 days.

How to Wax Cheese

Waxing "seals" cheeses that are to be cured, keeping out air. Although ordinary paraffin can be used, you'll get better results with a cheese wax available from cheese makers' supply companies.

1. Melt the wax in a double boiler to reduce the danger of fire. The wax should be fairly hot, but not so hot that it would burn a finger quickly dipped into it a little way.
2. Dip a cool, very dry cheese into the wax and immediately remove it. You don't want the cheese to warm up.
3. When this thin layer cools and solidifies, dip another section of the cheese in the same way.
4. Contue this process until you have an even coating of wax about 1/16 inch thick.

---

16. Wax the cheese. Age for up to 12 months at 40 to 50°F. You can serve it sooner, but age at least 60 days if you used raw milk. Turn once a day for the first several weeks, and once a week after that.

The possibilities for cheese making are endless. True, there are certain cheeses that can't be duplicated exactly, at home, including Limburger, Camembert, and others requiring special cultures that are closely guarded secrets. Some cheeses require special climates or particular caves for aging. But there are more than enough recipes and cultures to keep any cheese-loving goat owner happy, and very, very busy.

Using Whey

You don't have to discard the whey you ladle, pour, or drain off. You can make ricotta cheese with it (see note on page 217), or the dark brown cheese known as gjetost. Whey also makes good feed for chickens and pigs. Don't feed it to goats, either kids or adults; it will cause scours.

You can make an interesting "lemonade" with whey, as described in a nineteenth-century cookbook. Here's how: Strain 1 quart of whey. Add 6 tablespoons of sugar or honey, the juice of 2 lemons, and a dash of nutmeg or cinnamon. Serve chilled.

Yogurt

Yogurt has been eaten and cooked with for centuries throughout the Balkans, the Middle East, and India, where goats are the standard dairy animals. The word, which can also be spelled *yoghurt* (the older spelling), is in that form directly from the Turkish. Homemade yogurt is so superior to the supermarket variety that there's no comparison. There are several ways of making it, with everything from commercial yogurt makers, to heating pads, to solar energy. What's more, there are so many ways of using yogurt. Try it, for instance, as a substitute for sour cream or light cream. In Middle Eastern cooking, it's used as the liquid in many kinds of stews and soups. Goat-milk yogurt doesn't curdle when cooked, as cow-milk yogurt does.

Making Yogurt

You need a yogurt starter or culture. You can buy dried cultures, or you can use 1 cup of plain store-bought yogurt for your first batch and save a cup of your homemade product to start the next batch. This will die out after a while, and you'll have to use a fresh culture again.

First, warm the milk to 100 to 110°F. Add the yogurt culture. If you prefer a firmer product, add ½ cup of powdered milk to each 3½ cups of goat milk. (Use dry goat milk if you can find it.) Pour the warm milk plus culture into the cups of a commercial yogurt maker or your own home-made maker. The yogurt "cooks" in about 5 to 6 hours at a constant 100°F.

A yogurt maker will automatically keep the milk at the proper temperature. But a preheated thermos wrapped in towels to help hold the heat in works just fine. Or pour the warm milk and culture into a casserole dish, set it in a warm oven, and leave it overnight with the heat off. You can also make yogurt with a heating pad, or let it "cook" on the back of the old wood cookstove. On a warm sunny day, you can use sunlight: pour the milk and culture into a glass-covered container, and set it in the sun. (A solar oven, however, gets much too hot unless you tend it carefully.)

Problems You Might Encounter with Yogurt Making

First, don't use milk that contains antibiotics. These will kill the yogurt-making bacteria, and the milk won't "clabber," or curdle. It's these healthy lactobacilli, or lactic-acid bacteria, that actually produce the yogurt. You can get rid of the "bad," illness-causing bacteria by sterilizing all equipment used in the process.

Temperatures that are too high (above 115°F) will kill the lactic-acid bacteria, too. On the other hand, a temperature below 90°F will cause the bacteria to work too slowly. They haven't died, so the yogurt might still set if given more time. Disturbing or moving the yogurt during incubation can also hamper clabbering.

Moving the yogurt during incubation also creates too much whey. This happens as well if the temperature is just a *little* too high or if you incubate the yogurt too long.

If your yogurt is too sour, you may have incubated it too long or used too much starter. An "off" taste is usually attributed to the milk, unclean utensils, or old or contaminated culture.

Yogurt Cream Cheese

This is a delightful change from store-bought cream cheese. Pour yogurt into a cheesecloth-lined colander. One quart is a good measure to work with. Let it set all day or overnight, until the whey drains off and the yogurt becomes a light, creamy cheese. Use at once in place of cream cheese, or refrigerate it, in a covered container, for up to 1 week.

Try this for a breakfast treat, as they do in the Middle East, or for a healthful snack. Form the yogurt cheese into small balls. (If they're too soft, refrigerate them overnight.) Roll the balls in olive oil, and sprinkle with paprika.

You can use yogurt cream cheese in unbaked pies, too. (Maybe we should have started out with more than a quart!) Combine 3 cups of yogurt cream cheese with 3 tablespoons of honey, and 1 teaspoon of vanilla or grated orange peel, and stir until smooth. Pour into a 9-inch baked pie shell. Chill for 24 hours before serving. This pie is delicious as-is, but for a special treat you might want to add your favorite topping — perhaps strawberries, peaches, blueberries, or whatever is in season or suits your fancy.

Kefir

Kefir, a drink of Russian origin, is similar to yogurt, and it's incubated at room temperature (65–75°F) for 24 hours. In addition to lactic-acid bacteria, it contains a lactose-fermenting yeast that gives it a unique fizzy character.

There are two forms of kefir. One is made with a yogurtlike culture, the other from small, curdlike particles called kefir grains. These are added to the milk to clabber it and can then be strained out and reused indefinitely if properly cared for. The cultures are available by mail order.

Koumiss

You'll need a slight sense of adventure to try *koumiss* or kumiss, also called milk beer. It originated in central Asia and was traditionally made from mare's milk, which is said to give it a very high alcohol content. But, honestly, wouldn't you rather milk a *goat?* This is an Americanized version.

Thoroughly mix 1 quart of fresh-from-the-goat unchilled milk with 4 teaspoons of sugar and 1 teaspoon of dry yeast. Let stand uncovered in a warm place for 10 hours. Pour it back and forth between two pitchers until it's foamy and smooth. Then store it in a jar with a tight-fitting lid in a warm place for another 24 hours. Chill. Stir before serving.

Butter

Butter is difficult to make from goat milk, but only because of its so-called natural homogenization (see chapter 2). If you are fortunate enough to own a cream separator, set the cream-adjusting screw to the finest setting with goat milk.

If you don't have a cream separator, you'll have a hard time getting enough cream to warrant cranking up the churn. However, it is possible to make butter from goat milk without a separator. One method, which will separate at least some of the cream, is to leave the milk in a flat pan with as much surface as possible exposed to the air. You'll be able to skim off some of the cream in about 24 hours. The cream will keep for a week in the refrigerator, so you can skim the cream from daily milkings for up to a week to accumulate enough to make butter.

The cream should be "ripened," which will happen naturally if it stays in the refrigerator for a week. Otherwise, leave it at room temperature for a day.

Put the cream into the churn, but don't fill it more than half full. If you don't have a churn, use an electric mixer or a French whip, or just shake it in a canning jar with a tight lid, opening the lid once in a while to release the pressure.

The butter should begin coming together in about 20 minutes, in the form of pea-sized grains. Drain off the buttermilk. This isn't cultured buttermilk like you buy in the store, but it's drinkable and also good for cooking.

Now you have to work the butter. Using a spatula or wooden spoon, press the butter against the side of the bowl, and pour off the buttermilk that is pressed out.

Then wash the butter in cold water to get out any remaining traces of milk, which will cause the butter to spoil. Repeat the washing until the water comes off clear.

If you prefer salted butter, add salt to taste, and work it in.

You might be surprised to see that your goat butter is white. That's normal. If you want to make it yellow, use a special "Dandelion" butter coloring available from goat-supply houses. Regular food coloring won't stick to the butterfat.

Cultured Buttermilk

The culture in buttermilk sold in stores is created by adding certain organisms to regular sweet milk and cultivating, or culturing, them.

The simplest way to culture buttermilk at home is to warm 1 quart of fresh goat milk to 72°F, and then add 2 tablespoons of store-bought buttermilk. Stir; then let sit at room temperature for 8 to 12 hours, or until it's as thick as you like it. Refrigerate. You can use 2 tablespoons of this batch to start the next batch. Buttermilk is also used as a culture in some cheeses, as in the Gouda recipe on page 222.

How Else Can I Use Goat Milk?

Whenever a recipe or formula calls for milk, you can use goat milk. Conversely, any recipe specifying goat milk will work just as well with cow milk. There is no need to search for special recipes. When you have fresh, wholesome, delicious goat milk, that's the only kind to consider using, whatever you're making.

RECIPES
FOR
GOAT PRODUCTS

People who have fresh, delicious milk from their own goats probably drink more milk than people who buy cow milk in stores. Even then, they frequently have a surplus and search for recipes in which milk is a major ingredient.

Making cheese and the other products covered in chapter 16 can use large quantities of milk, but this chapter inludes more ideas that will help you keep smaller surpluses from going to waste. And remember, you don't need special recipes for goat milk. Use goat milk in any recipe calling for milk.

Similarly, you don't need special recipes for chevon. It can be used in any dish calling for beef, pork, or certainly lamb. It can replace beef or venison (but not pork) in sausage.

Make full use of the goat products that are harvested from your backyard dairy. These family favorites that readers have shared with *Countryside* magazine over the years should help you get started.

Milk Soup

When your goats are producing plenty of milk and your hens are providing
more eggs than you need for your usual cooking and baking,
try this unusal "waste-not-want-not" dish.

    2 quarts goat milk
    ½ teaspoon powdered cinnamon
    1 teaspoon salt
    1 tablespoon powdered sugar
    4 thin bread slices
    6 egg yolks

1. Boil the milk with the cinnamon, salt, and sugar.

2. Lay the bread in a deep dish, pour a little of the milk over it, and keep hot, but without burning it.

3. Beat the egg yolks, and add them to the milk. Stir the mixture over low heat until thickened. Do not let it curdle.

4. Pour the egg-milk mixture over the bread, and serve.

Devonshire Cream

In Devonshire, England, in the nineteenth century, this cream was used
to make a very firm butter. But it was also considered a gourmet item
in London, where it was eaten with fresh fruit. You can serve it with
fresh berries or with warm scones and strawberry jam.

1. Let milk stand 24 hours in the winter, 12 hours when the weather is warm.

2. Set the pan on the stove over very low heat, and heat the milk until it is quite hot. Don't let it boil; the longer the heating takes, the better. When it's ready, there will be thick undulations on the surface, and small rings will appear.

3. Set the pan in a cool place for 1 day.

4. Skim off the cream, and serve.

Cream Cheese

⅛ cake compressed yeast
11 cups goat milk (amount depends on how long
    the culture takes; see step 3)
  Salt, sugar to taste

1. Make a culture: Add the yeast to 1 cup of warm goat milk. Let it stand in a warm room for 24 hours.

2. Pour off half of the culture, and add 1 cup more of warm goat milk.

3. Repeat this process, pouring off half the amount and adding 1 more cup of warm milk, every day for about a week, until the yeast flavor disappears.

4. Add the culture to 2 cups of warm goat milk. Let it sit for 24 hours.

5. Heat the resulting curd over hot, not boiling, water for about 30 minutes, until it is firm.

6. Press the curd gently in cheesecloth to remove the whey.

7. Add salt and sugar to taste. Keep refrigerated until served.

Yogurt Variations

For a basic yogurt recipe, see page 225.

- **Bavarian cream.** For each 2 cups of yogurt, add cool but unset gelatin-dessert mix (your choice of flavors), made double strength.

- **Sherbet on a stick.** Stir frozen juice concentrate (to taste) into yogurt, and spoon into small plastic cups. Insert a plastic spoon into the center, and freeze. To serve, unmold and use the spoons like handles.

- **Sour cream.** Spoon any amount of yogurt onto a clean cloth, draw up the corners, and hang it to drain for 3 hours, or until it's as firm as you want it. Use in any recipe calling for sour cream, but be warned that it breaks down under heat. To use it in beef Stroganoff or other cooked dishes, add cornstarch or flour as a stabilizer, and heat and stir gently.

Sweet Cheese

This is a delicious mild cheese.

1 gallon milk
1 pint buttermilk
3 eggs, well beaten

1. Bring the milk to a boil.

2. Add the buttermilk and the eggs. Stir gently.

3. When the curd separates, drain, and press.

Goat-Milk Fudge

You'll need a candy thermometer for this fudge.

2 cups sugar
2½ squares baking chocolate
1 cup goat milk
¼ teaspoon salt
1 cup nuts, chopped
1 teaspoon vanilla
1 tablespoon goat butter (see page 227)

1. Mix the sugar, chocolate, goat milk, and salt in a heavy saucepan. Cook over medium heat, stirring constantly.

2. Bring to 236°F on a candy thermometer, or to soft-ball stage (a few drops dribbled into cold water can be formed into a soft ball).

3. Remove from heat, add the nuts, vanilla, and goat butter. Beat until thick and creamy.

4. Pour into a buttered dish and cool. Cut into squares.

Goat-Milk Pudding

You can also fill cream pies with this pudding.

2½ cups goat milk
½ cup sugar (use brown sugar for a butterscotch flavor)
Pinch salt
1 egg
4 tablespoons cornstarch
1 tablespoon goat butter (see page 227)
1 teaspoon vanilla
2 heaping tablespoons sweetened cocoa powder
(for a chocolate pudding)

1. Mix 2 cups of the goat milk, the sugar, and the salt in a heavy saucepan. Heat slowly.

2. While the milk mixture is heating, beat the egg. Add to milk mixture, and bring to the scalding point, stirring constantly.

3. Dissolve the cornstarch in the remaining ½ cup milk, and add to the scalding milk, again stirring constantly. Stir until thickened, and remove from heat.

4. Add the goat butter and vanilla.

5. For flavored puddings, mix in the cocoa with the sugar before adding the milk, or substitute brown sugar for white for butterscotch flavor.

Some Tips on Making Ice Cream

♦ You can increase these recipes, but remember that ice cream expands as it freezes. Don't fill the container more than three-fourths full.

♦ Use cream that's a day old for a finer-grained product than one using fresh cream.

♦ Prepare the mixture a day ahead of time for a smoother product. This also increases the yield.

Vanilla Ice Cream I

This recipe is easy, but it doesn't make very much.

2 eggs, separated	⅝ cup heavy cream
Scant ½ cup powdered sugar	⅝ cup goat milk
A few drops vanilla extract	

1. Beat the egg yolks, sugar, and vanilla in a bowl.

2. Meanwhile, bring the milk to a boil. Pour it over the egg-sugar mixture, stirring constantly. Cool, then refrigerate until quite cold.

3. Whisk the egg whites until stiff.

4. Lightly whip the cream.

5. Fold the egg whites and cream into the cold egg-sugar-milk mixture. Whisk well.

6. Pour the mixture into shallow trays, and freeze until slushy.

7. Return the mixture to the bowl, and whisk again.

8. Pour back into trays, and freeze again.

9. When frozen, refrigerate for 30 minutes or so to soften it slightly.

Vanilla Ice Cream II

This recipe is similar to the first one, but it doesn't use eggs and is less rich. It makes about 1½ quarts.

4 cups cream	⅛ teaspoon salt
1 cup sugar	1½ teaspoons vanilla

1. Warm, but do not boil, 1 cup of the cream over low heat.

2. Stir in the sugar and salt until they're dissolved. Chill overnight.

3. Add the remaining 3 cups of cream and the vanilla, pour into the canister of an ice cream freezer, and proceed according to manufacturer's instructions.

Vanilla Ice Cream III

*This recipe requires a little more work than the others.
But if you enjoy the premium, high-priced,
store-bought ice creams, it's worth the extra trouble.*

    4 eggs, lightly beaten
 1½ cups sugar
    ½ teaspoon salt
    2 cups goat milk
    2 cups light cream
    1 tablespoon vanilla powder*
    4 cups heavy cream, well chilled

1. Combine the eggs, sugar, and salt in the top of a double boiler.

2. Whisk in the milk and light cream, and cook over simmering water, stirring constantly. When the mixture thickens slightly, remove it from the heat.

3. Add the vanilla powder, straining it through a large mesh sieve to remove any portion of the bean that is not finely ground.

4. Stir thoroughly, and refrigerate, preferably overnight but for at least several hours.

5. Just before you're ready to start cranking, remove the cold custard from the refrigerator, and blend in the cold heavy cream. Pour the mixture into the canister of the ice cream freezer, and crank.

*Note: Vanilla beans are expensive (and hard to find in some places), but they're a real treat. To make the powder, grind several dried vanilla beans in a spice mill. One 4-inch bean will make about 2 teaspoons of powder. The tiny specks of vanilla will show, but in ice cream, that's wonderful.

Cooking Chevon

Milk is the most important reason for having goats, but they provide more than dairy products. Goat meat, or *chevon*, is excellent! Use it just like beef, lamb, or venison, or try some of the ethnic dishes that are popular now from the Caribbean, Asia, and Africa, where goat meat is a staple.

Chevon, like buffalo meat, has a very low fat content. Low-fat meat of this type should be cooked slowly, and at low temperatures, with moist heat. Rapid cooking at high temperatures, without added moisture, will result in a tough, dry, flavorless product. Note that this is just the opposite of tender, juicy, flavorful meat, which chevon certainly can be if properly prepared.

Chevon does have a distinctive flavor and aroma, quite unlike beef, pork or venison. Many people value this difference and want recipes that preserve and enhance it. This might involve simply browning the meat in olive oil and roasting it with salt and pepper, so the natural flavor isn't masked. To discover why goat meat is so popular in Greece, add some oregano, garlic, and lemon juice. In other places, goat meat curry is preferred.

Some people don't appreciate the flavor, and if they have to eat chevon only because they have excess goats, they prefer recipes that hide or eliminate the taste. This is easily done with marinades, or cooking in any number of sauces, especially those containing tomato. Simmer it slowly with fresh or canned tomato chunks or sauce, fresh or dried celery leaves, garlic if you like that, oregano, cilantro, Worcestershire sauce, lemon juice, or anything else you fancy. Marinate the meat in cold black coffee. Or grind the meat and mix it with ground beef or pork.

If you have cooking experience and creativity, you won't have any trouble preparing delicious meals with chevon. If you need some help, look for recipes in Mexican, Greek, and other ethnic cookbooks.

But also remember, you don't need any special recipes. In fact, in some households with fussy eaters it might be wise to get your family accustomed to chevon by starting out with familiar dishes, especially stews, casseroles, or chili, that involve a variety of ingredients and spices, slow cooking, and moist heat.

Chevon Chili

This recipe comes from Frank Pinkerton, "the goat man"
of Langston University in Oklahoma. Frank does not recommend
adding beans to this recipe. Serve pinto beans as a side dish.

MAKES 14 8-OUNCE SERVINGS

    2 cups onions, chopped
    2 tablespoons olive oil
    1 tablespoon ground oregano
    2 tablespoons ground cumin
    1 teaspoon garlic powder
    1 tablespoon salt
    3 pounds ground or cubed goat meat
    ½ cup chili powder, or to taste
    ½ cup flour
    8 cups boiling water

1. Sauté the onions in the oil in a cast-iron pot, if you have one, or in a heavy Dutch oven or stockpot.

2. Add all spices except chili powder. Stir occasionally.

3. When the onions are almost clear, add the meat. Simmer until gray.

4. Add the chili powder and flour, and stir vigorously to thoroughly blend everything.

5. While stirring, add the boiling water, and bring the entire mixture to a boil.

6. Simmer for not more than 1 hour. Add other seasonings, such as cayenne or hot peppers, at this time, if desired.

Try Chevon in All Your Lamb Recipes

Because chevon has a lamblike flavor, the cuts are similar, and, like lamb, chevon is somewhat dry and quite lean, any recipes for lamb are equally good with chevon, whether they call for roasting, baking, or grilling or for ground meat. Try ground chevon in homemade sausage.

Chevon Stew

Does anyone need a recipe for stew? Well, probably at first, but after that if it's not instinctive you aren't making stew. (Please forgive me, but next to goats, cooking from scratch, without recipes, is one of my greatest passions and pleasures. And for homesteaders like me, making something out of nothing and eating with the seasons is a virtue, a necessity, and a point of pride.) Here's how I make stew.

You know stew is on the menu when all the "fancier" cuts of meat have been used, the fresh summer vegetables are gone, and some of those in the root cellar are calling for attention. Or maybe it's just a cold wintry morning and you feel like simmering a kettle of stew on the woodstove all day. Take inventory and gather your ingredients.

The meat is paramount, of course. It should be cubed, in 1-inch squares or smaller. Or larger, if you like.

Brown it in olive oil. Or bacon grease. Or butter, which tenderizes it. Even canola or vegetable oil will work, but being a peasant gourmet I really prefer olive oil or butter.

Add onions. Chopped, sliced, diced, minced, whatever you feel like. Even small whole ones, if you have them or like them. As many as you want or don't want. Garlic, too. Homegrown. Sliced, diced, minced, or juice. Even whole cloves. If you don't grow or buy garlic cloves, powder is okay.

Brown (sear) the meat, because it won't brown after liquid is added. Yes, this breaks the low-heat rule. If you insist on following rules, don't brown it. Just heat it to a sickly pale gray. (There are no rules for stew.)

Sprinkle some Worcestershire sauce over the meat and stir it, then scatter in about a tablespoon of flour (be sure there's enough juice/oil at this point) and stir it well to coat the meat pieces. More flour, or less, won't hurt anything, but this step determines how thick the gravy will be. You can thicken it more later if you want to though. Or thin it, although then you could end up with soupy stew. Which is good, too.

The floured meat will be a thick pasty glob. Add water to thin it, and stir it briskly to make a gravy. How much water? As much as you want, or just enough to make it look like stew.

Sprinkle this with crushed homegrown oregano leaves, basil, parsley, cilantro, whatever, but certainly celery leaves. You can use the stems, too, of course, but the leaves have the most flavor, and in our garden we get more leaves than stems.

Then add cut-up potatoes — as many as you think looks good, in whatever shape or size appeals to you. Again, 1-inch (or less) cubes work well. Or use small whole ones. Carrots, definitely. And turnips, Jerusalem artichokes, rutabagas, parsnips, celeriac, corn, peas, green beans — whatever you have that wants eating and will fit in your stew pot.

I didn't mention green peppers (or red or yellow ones) or mushrooms (wild, from your inoculated logs, or store-bought), but by now you should be getting the idea.

Start this in the morning, let it simmer on the woodstove all day, serve it with thick slices of fresh homemade bread or scones with homemade goat butter when the crew comes in from barn chores on a frosty winter evening, and they'll swear that chevon is as good as it gets.

Barbecue Sauce for Chevon Ribs

Chevon chops tend to be small, and I generally just leave them with the ribs. Barbecued ribs are popular at my house. Here's a tasty sauce.

  ½ cup onion, chopped
   1 clove garlic, crushed
   1 tablespoon fat or drippings
  ½ cup water
   1 tablespoon vinegar
   1 tablespoon Worcestershire sauce
  ¼ cup lemon juice
   2 tablespoons brown sugar
   1 cup chili sauce (or home-seasoned tomato sauce)
  ½ teaspoon salt
  ¼ teaspoon paprika
    Thyme, oregano, dry mustard, hot peppers, or
      pepper sauce (optional)

1. Sauté together the onion and garlic in the fat until tender.

2. Combine all the other ingredients, and simmer for 20 minutes.

3. Pour the sauce over well-browned ribs, and bake or grill until tender.

Marinating Chevon

Any meat can be marinated. The essential marinade ingredient is an acidic liquid, such as wine, vinegar, lemon juice, tomato juice, or cold coffee. The acid tenderizes the meat and enhances the flavor.

There are, of course, many marinades. If you're in a hurry, just splash some red wine and water on the meat in a dish, and sprinkle on whatever herbs and spices you like. (Garlic and oregano are the standards at our house.) Try the following suggestion if you have time for something a little more elaborate.

Marinated Chevon Steak

This recipe is for about 3 pounds of chevon steak.

Cider vinegar and water (half and half) to cover meat
¼ cup honey
1 large onion, sliced
2 teaspoons salt
¼ teaspoon paprika
4 cloves, whole
2 bay leaves
1 teaspoon oregano
½ teaspoon dry mustard
1 cup thick sour cream

1. Arrange the steaks in a shallow glass dish. Cover with the water and vinegar.

2. Add everything else except the sour cream.

3. Marinate in the refrigerator for 2 days.

4. Remove the meat, pat it dry, and dredge it in flour and butter. Brown the meat.

5. Add the marinade, and simmer gently until tender.

6. Stir in the sour cream, and serve.

A Source for the Home Sausage Maker

One source for everything you'll need to make sausage (except the meat, of course) is The Sausage Maker, Inc., 1500 Clinton Street, Bldg. 123, Dept. 20009-0f, Buffalo NY 14206 (www.making sausage.com). Ask for the company's classic publication, *Great Sausage Recipes.* It's full of treasures for the at-home sausage maker.

Making Sausage with Chevon

Sausage making is another one of those projects that can convert a backyard dairy into a homestead or become a fascinating hobby in itself. As with cheeses, there are hundreds of sausage recipes. They all have a common starting point; they can very simple or very complex; and they can easily be made in your kitchen with very little special equipment.

D. L. Salsbury, a veterinarian and an avid sausage maker who has shared many recipes with *Countryside* readers, likes to point out that while some people think sausage making has to be complicated, if you have ever made meat loaf you have made sausage. It's the same basic process.

Any butchering project will end up with bits and pieces of meat that might not be suitable even for stews. They're ideal for sausage. Older cull animals might best be utilized by using the entire carcass for sausage. But any meat, even the best cuts, can be included in sausage.

When making sausage with chevon, you must add at least some pork fat because chevon is lean and dry. It's the same with venison, so venison sausage recipes, which are quite common, work very well with chevon. You might even be able to find premixed spices and curing agents for venison sausage in your local supermarket, at least during hunting season.

For starters, try a simple recipe, one that doesn't require casings and stuffing, smoking, or aging. Here's an example.

Chevon Sausage

You can alter this recipe easily to suit your own tastes. You might want to add garlic, onions, your favorite herbs, or lemon juice. You can eliminate the potato. (The egg is essential; it acts as a binder.) Try adding finely diced green or red peppers, or hot peppers if you enjoy those. Be creative! Have fun!

MAKES 4 SERVINGS

1 pound chevon, ground
1 medium potato, boiled, peeled, and mashed
½ cup spinach, chard, or dandelion greens, parboiled, drained, pressed dry, and pureed
1 egg
2 tablespoons Parmesan cheese, grated

Salt and pepper, to taste
¼ cup bread crumbs, plus extra as needed
¼ cup flour
4 tablespoons butter
1 tablespoon olive oil
½ cup white wine
2 cups beef stock

1. With clean hands (no rings), mix the meat, potato, greens, egg, cheese, salt, and pepper in a bowl.

2. Add just enough bread crumbs, if necessary, to make the mixture hold together when formed into a ball. Mash it well, squeezing it between your fingers.

3. Then shape the meat mixture into a loaf or cylinder. Combine the flour with the ¼ cup of bread crumbs, and coat the meat with this mixture.

4. Brown the meat on all sides in the butter and oil.

5. Add the wine.

6. When the wine has almost evaporated, add the stock, cover, and simmer over low heat, turning occasionally, for 1 hour, or until the inside is done.

7. Remove the meat from the pan, and boil down the liquid until it's thickened.

8. Slice the loaf, and pour the juice over the slices to serve.

Soap Making with Goat Milk

Like cheese making and sausage making, soap making can become a plea-surable hobby. You won't use up much milk making soap, although I do know someone who got into goats as the result of obsessive soap making.

Milk is used in soap recipes because its casein thickens the soap and adds texture. Goat-milk soap is popular because it's white; soap made with cow milk turns orange. The following recipe makes a very nice soap.

Goat-Milk Soap

```
    1 can lye
    3 pints goat milk
 5½ pounds goat fat (or other animal fat), clarified and
      lukewarm
    4 heaping teaspoons borax
    2 cups oatmeal, finely ground
    2 ounces glycerin
```

1. In a large kettle over low heat, add the contents of the can of lye to the goat milk. (Be sure to follow the directions and observe the warnings on the lye-can label.) Stir with a wooden spoon.

2. When the mixture is warm (don't touch the lye solution; just feel the outside of the pot), pour the fat into it as you stir.

3. Add the borax and oatmeal.

4. Add the glycerin, and stir 15 to 30 minutes, or until the mixture starts to harden.

5. Pour the mixture into molds (plastic-foam drinking cups make fine molds), or using rubber gloves, shape into balls.

6. Let the soap ripen for 3 weeks or more in a moderately cool, dry, airy place. The older the soap, the better. It might have a peculiar odor at first, but this will disappear with age.

Other Ideas

Though it's now out-of-print, *Caprine Cooking* by Mary Jane Toth has almost 500 pages of recipes, including several for soap. Check your library or local used-books store for this gem.

What's Next?

We haven't mentioned making useful and beautiful leather and fur from goat hides. (Tanned kid skins are very much like fur.) You can make candles from goat tallow. When you become accustomed to dining on chevon and making sausage, there will be no such thing as "unwanted" kids. When you get really good at making cheese, yogurt, butter, and ice cream, and you use fresh delicious goat milk in custards, baking, and other cooking, you might find that you don't have enough milk left to drink, which would be a tragedy.

The solution? Simple! Go back to the beginning of this book. Then get more goats!

APPENDIXES

A. Milking Through

Ordinarily, goats are bred annually, to kid once a year. That's because a doe produces milk to feed her kids and, at least in theory, she needs to give birth on a regular basis in order to produce a reasonable quantity of milk. This can be seen in the lactation curve: the amount of milk produced daily increases rapidly after parturition, then slowly drops off, reflecting the natural milk requirements of the young.

A 305-day lactation is considered standard, but not because its the norm. Milking records are based on a 305-day lactation, assuming 365 days between birthings and a 60-day dry period for the doe to rest and replenish her body reserves. Yet, one study of goats on official test showed that only about one-third milked for 305 days; the rest were dry before that.

In spite of this, we're hearing more and more about goats "milking through." This refers to does that continue to produce respectable quantities of milk for 2 or 3 years, or even more, without being rebred. For many home dairies, this is an attractive option.

The main problem is finding a doe that is capable of such production. About two-thirds of the goats on official test don't produce milk for even 305 days, much less a year or 2 or 3. And if we assume that goats on test are likely to be better than average and most beginners will start with an average animal, the chance that your first goat will be capable of milking through is pretty slim.

How will you know if you have a goat with this potential? If she continues to produce well even when she would normally be bred. How much milk that it is up to you, but most home dairies would be satisfied with 4 to 5 pounds a day. While a commercial dairy needs a certain amount of milk per doe per day to be profitable, a home dairy might be satisfied with just a few glasses a day, when the alternative is buying cow milk.

The Advantages

A family goat that milks through offers many advantages.

One is that you'll have milk year-round. The "normal" system of drying off after 305 days leaves you with the unpleasant situation of buying feed for the goat and milk for your family, for 2 months of the year! The option is having several goats freshening at staggered intervals, which, among other things, means having a flood of milk at certain times, and isn't always convenient, or possible. (It might involve winter lighting schemes or hormone implants.)

Another big advantage for many people with just a few milkers is that they don't need a buck or don't have to transport the does for breeding. This reduces costs, time, and labor.

A third advantage won't apply to as many goatkeepers as the first two, and might not be as readily apparent, but it can be significant. That is unwanted kids. For many people, kids are the most delightful part of raising goats. Why would anybody not want them? And on a homestead, if they're not needed as milkers, the meat is an important by-product of the dairy enterprise.

But raising kids requires a great deal of time and effort. There is also the expense — of milk lost to the household or of milk replacer. And let's face it, many goat raisers can't even think of butchering and eating kids. But if your little herd doubles or triples in size every year, something has to give. (Do the math. If a doe has doe twins once a year, and her kids and their kids have doe twins, you'll go from 1 to 3 to 6 to 12 to 24 to 48 goats in just 5 years! Not all kids are females, naturally, but you will get "compound interest" from your goats.) You can't keep them all. And if disposing of them is a problem, milking through is an elegant solution.

Other reasons for milking through are more subtle. One is that kidding complications, such as ketosis, are a leading cause of death in many herds. When kidding is reduced to every 2 or 3 years, the goats live longer.

Remember, not all goats are persistent milkers. Saanens are most likely to milk through, but individuals of other breeds will too, and, of course not all Saanens will.

Like many other management practices, milking through depends on your specific animals and your own needs and preferences.

Not all goats will milk through. Not all goat raisers want theirs to. But for some, the advantages are attractive.

Tips for Milking Through

- ◆ The doe must be in excellent health and condition and free of parasites. Good feed is essential.

- ◆ If there are no bucks around, the doe will be less likely to exhibit strong heat periods, which can contribute to lower milk production.

- ◆ And since shorter days contribute to a drop in milk production (referred to as photoperiod sensitivity), keeping lights on in the barn can help. Sixteen hours a day is the ideal.

B. Where Milk Comes From

What makes a goat "let down" her milk? Why does she sometimes "hold back" milk? What accounts for the lactation curve, and what makes a goat dry off? In brief, where does milk come from?

As a goat owner, you're almost certain to ponder these questions while milking, sooner or later. Here are some of the answers.

Like all mammals, female goats produce milk for the purpose of feeding their young. Thus, the doe must be bred and give birth before she will lactate (with certain abnormal exceptions), and lactation stops naturally as kids are weaned. This period of milk production can be extended somewhat, but not indefinitely. Eventually the doe must be bred before she will start producing milk again.

Activity in the Udder

The goat has two mammary glands, which collectively are called the *udder*.

Probably less than 50 percent of the milk an animal produces can be contained in the natural storage area of the udder. The balance is accommodated only by the stretching of the udder. In some cases, this can cause the ligaments that suspend the udder to become permanently lengthened, causing the udder to "break away" from the body, resulting in what is called a *pendulous udder*.

A good udder is capacious, it has a relatively level floor, and is strongly attached. It has plenty of glandular tissue, but a minimum of connective and almost no fatty tissue. After milking, the normal, high-quality udder feels soft and pliable, with no lumps or knots that would indicate scar tissue resulting from injury or disease.

The udder is divided into right and left halves by a heavy membrane. The milk produced in each half can be removed only from the teat of that half.

The Teat, Lobes, and Alveoli

Why are some goats hard to milk and others "leakers?" Let's take a closer look at the mammary system, starting with the teat, and working back to the origin of the milk itself.

The teat has an opening at the end, known as the *streak canal*, which is surrounded by sphincter muscles. These muscles prevent the milk from flowing out and their strength determines how hard or easy it is to milk the animal. Goats with very strong sphincters might be hard to milk, while those with weak sphincters can actually drip milk. (We'll see that these sphincters are affected by hormones resulting from premilking routines and other factors and why it's important to avoid changing or upsetting those routines, among other things.)

Following the path of the milk back to its origins, we see where the teat widens into its *cistern*, the final temporary storage area before milking or suckling. Beyond that is the *gland cistern*, which is in the udder; and beyond that are large ducts which branch through all parts of the udder to collect and transport the milk toward the teat.

Each duct drains a single *lobe*, which further branches into *lobules*, which is where milk production actually takes place.

These lobules are composed of many tiny, hollow spheres called *alveoli*, which are said to resemble bunches of grapes — very tiny grapes. One cubic centimeter contains about 60,000 alveoli. Milk is secreted in the cells of each alveolus.

Each alveolus is surrounded by muscle fibers that balloon out as milk is secreted. When the animal is properly stimulated to let down her milk, these microscopic muscles contract, forcing out the milk. Inadequate or improper stimulation can therefore result in "hard milking."

Blood Supply

Blood supply to the udder is important in the making of milk. For each unit of milk secreted, anywhere from 300 to 500 times as much blood passes through the udder!

Blood enters through the base of each half of the udder through the pudic artery. It returns to the heart for oxygenation by one of three passages. One of these is the subcutaneous abdominal vein, often called the milk vein. Many milkers believe the size and prominence of the milk vein is an indication of milking ability. However, experimenters have tied off these veins, and milk production wasn't appreciably affected.

Nevertheless, milk is made largely from constituents of blood. A tiny network of capillaries surrounds each alveolus. These capillaries carry blood to the base of the cells that line the alveolus. Milk is made when materials in the blood passing through these capillaries are taken up by the milk-making cells. The process involves, partly, filtration of certain constituents in the blood itself, and partly synthesis of constituents by cellular metabolism.

Scientists have been able to determine the blood constituents used in milk production by examining blood before it enters and after it leaves the udder. Water, the largest component of milk (87 percent by weight), is filtered from the blood, as are milk's vitamins and minerals. The lactose in milk (its principal carbohydrate) is synthesized from blood glucose.

On the other hand, about 75 percent of the fat in milk is synthesized in the mammary gland. Most of this comes from acetate, which explains why animals on high-grain and low-forage diets often produce milk with a lower fat content. This kind of diet results in reduced production of acetate in the rumen.

Secretion of the Milk

Now it's milking time. The cells of the alveoli have been busy making milk by filtering blood and synthesizing other constituents. These tiny cells have lengthened as the milk accumulated. When filled, the cells ruptured, pouring their contents into the lumen, the cavity of the alveolus. This caused the mammary glands to become saturated with milk, like a sponge.

Incidentally, all this activity is most rapid immediately after milking. As milk is secreted in the cells and collected, increasing pressure in the mammary system slows the secretion-discharge cycles. Each hour after milking, milk production decreases by 90 to 95 percent. At a certain point (technically, at a pressure of 30 to 40 millimeters in mercury, which is roughly equal to capillary pressure and about one-fourth of systemic blood pressure), milk secretion is reduced appreciably, or stops entirely. If the animal isn't milked, the milk starts to be resorbed into the bloodstream.

Also of practical interest, when the udder is full, the secreting cells are unable to rupture because of the pressure. Therefore, only that part of the milk that can pass through the semipermeable cell membrane can be discharged. The milk fat is not discharged. This lower fat milk dilutes the milk previously secreted. That's why when the interval between milkings lengthens, the fat content of the milk decreases.

This also explains several other important phenomena, such as why milking three times a day results in more milk than milking twice a day. Since secretion is highest soon after milking, with three milkings it progresses at a faster rate for more hours of each day. This is also the reason the strippings are higher in fat than the first milk drawn; the cells with accumulated milk fat are able to discharge the fat globules when udder pressure has been reduced. And in addition, it helps explain why the milk-fat percentage is usually higher with lower producing animals or those whose production is declining in late lactation.

Now the stage is set. The goat is almost ready for you to start milking. Almost, but not quite! Coaxing milk from the udder involves more than just squeezing teats. What are required now are the only stimulators of lactation: hormones.

Hormones

Hormones are closely related to milk production in many ways, one of the most obvious being the development of the udder itself. The udder is, after all, a secondary sex characteristic. Its very existence and function are the results of hormonal activity.

Six hormones are important in the intensity of lactation.

1. Prolactin is secreted by the anterior pituitary gland and, in mammals, stimulates the initiation of lactation. It also increases the activity of the enzymes that are essential to the work of the epithelial cells (in the alveoli), which convert blood constituents to milk.

2. **Thyroxine** is secreted by the thyroid gland. Cows experimentally deprived of this product of thyroxine have gone down in milk production by as much as 75 percent. Other research has shown that thyroxine secretion increases in the fall and winter, and decreases in spring and summer. This is said to partially explain why milk production decreases in hot weather.

3. **Somatotropin,** secreted by the anterior pituitary gland, is also known as bovine growth hormone, or BGH. It regulates growth in young animals but also influences milk secretion by increasing the availability of blood amino acids, fats and sugars for use by the mammary gland cells in milk synthesis.

4. **Parathyroid hormone,** secreted by the parathyroid gland, regulates blood levels of calcium and phosphorus, which are major constituents of milk. As such, it also plays a role in milk fever (see chapter 8). When an animal freshens and starts producing milk, the mammary glands rapidly withdraw calcium and phosphorus from the blood. Without proper nutrition and without parathyroid hormone, the animal can develop milk fever. The practice of feeding high levels of vitamin D before freshening is related to this hormone.

5. **Adrenal hormones,** products of the adrenal glands, work both ways — small amounts are essential to milk production, but larger amounts will depress it. This is why it's important not to startle or frighten dairy animals. When disturbed, adrenaline is secreted to overcome the stress of the moment, but it decreases milk secretion.

6. **Oxytocin** is secreted by the hypothalamus and works with prolactin. (It also induces expulsion of the egg in the hen, and is used to induce active labor in women or to cause contraction of the uterus after delivery of the placenta.)

Stimulation of Hormones

A goat doesn't voluntarily "hold back" her milk. But she does have to be properly stimulated. When she is, milk is suddenly expelled from the alveoli into the large ducts and udder cisterns. Then, and only then, can it be removed.

The natural stimulus is nursing. However, manual massage of the teats and udder, performed while washing them, has the same effect. In addition, sights, sounds, and smells have an effect on milk letdown. Your arrival at the barn, turning on the lights, feeding: All the routines of milking are signals for the milk to start pumping. This is one of the reasons many goats will produce less when moved to a new home.

Then oxytocin is poured into the bloodstream, reaching the udder in 30 to 40 seconds. This causes the cells to contract, squeezing milk from the alveoli. Milk pressure in the cistern is almost doubled. However, this lasts for only 8 to 12 minutes. (This backs up the belief that to get the most milk, you have to milk fast.)

Oxytocin can be neutralized by adrenaline. This hormone increases blood pressure, heart rate, and cardiac output. It also causes the tiny arteries and capillaries of the udder to constrict and, thus, prevents oxytocin in the blood from reaching its destination. And adrenaline remains in the blood longer than oxytocin.

Adrenaline, of course, is released when the animal is in pain, frightened, irritated, startled by a loud noise, or otherwise bothered. Small wonder that the first-time milker who pains, frightens, irritates, startles — and probably embarrasses the goat — gets so little milk!

Drying Off, or Involution

You never get the "last drop" of milk from an udder. Normally, 10 to 25 percent remains even after stripping.

But if the milk isn't removed from the udder, the milk already there will be resorbed and the cells will quit producing more. Even just incomplete milking causes pressure to build more rapidly, and less milk is secreted between milkings. Eventually the secretion is impaired, and the animal dries off.

A goat can be dried off by milking on alternate days, or by stopping milking altogether. Quitting cold turkey is preferred.

Other Factors Affecting Lactation

As we've seen, hormones play a major role in milk production. But they aren't the only factors. There are many others.

1. Genetics. Genetics affects milk production. If the animal hasn't inherited the potential to produce milk, including the capability to produce the needed enzymes, she won't produce milk.

2. Secreting tissue. The animal must have enough secreting tissue. A small udder has proportionately less of this tissue, and it also has less capacity for storing the milk secreted. (This doesn't necessarily mean a large udder is better, but an udder should have *capacity*.)

3. Stage of lactation. Milk production generally peaks within 1 to 2 months of freshening, then drops off. The rate of this drop-off, or persistence, has a marked effect on annual production. Persistent milkers might drop 2 or 3 percent a month; others drop off much faster. Easy milkers, those that milk out more rapidly, are usually more persistent.

4. Frequency of milking. As we've seen, frequent milking tends to lengthen lactations.

5. Stage of pregnancy. Females in later stages of pregnancy generally drop off in milk production quite rapidly, because nutrients are diverted from the mammary gland to the uterus, for growth and maintenance of the fetus.

6. Age. Milk production increases with age, up to a point, but then drops off with advancing years.

7. Estrous cycle. Milk production drops when an animal is in heat.

8. Health. As might be expected, any disease can reduce milk production. Diseases can slow the circulation of blood to the udder, and again for reasons already given, that affects milk secretion.

9. Feed and nutrition. In the hierarchy that nature has wisely established, a body's first responsibility is survival. Maintenance comes before milk production.

10. Temperature. High temperatures decrease milk production. This has several explanations, including depressed appetites, reduced thyroid secretion, and others. Optimum temperatures seem to be between 50 and 80 degrees.

11. Milking routine. If you don't properly stimulate your goat, you won't get all the milk possible, and that in turn will decrease production at future milkings.

Milking isn't just a chore; it's participating in a marvel of nature. And lucky people who have goats get to do it twice a day!

C. Somatic Cell Counts

As a beginning goat milker and goat-milk drinker, our first interests are likely to concern quantity, flavor, and basic food safety (such as animal health factors associated with milk, and sanitation procedures). You could milk goats for a long time without even hearing about somatic cell counts (SCCs). Sooner or later you will — and you'll probably wonder what all the fuss is about. Here's a simplified explanation.

Somatic cells are white blood cells that are routinely sloughed off into the milk. Excessive amounts can indicate a problem such as mastitis or that infectious bacteria might be present in the milk. "Counting" these cells in a milliliter of milk is the basis for mastitis tests. For commercial dairies, the SCC must be below certain levels. This is where the trouble and confusion began.

It has been known for many years that goats have higher SCCs than cows, even when further tests show no mastitis or infectious bacteria is involved. Yet, common mastitis tests (such as the California mastitis test, or CMT) were designed for cows. This becomes a particularly vexing problem for commercial goat dairies that must meet regulatory SCC standards.

Until recently, the federal limit for SCCs in grade A milk, cow or goat, was 1 million/mL. It was lowered to 750,000 for cows, remaining at 1 million for goats.

However, an average cow dairy has an SCC of around 300,000 and well-managed herds can come in at under 100,000. Not so with goats, where even a well-managed herd is likely to have an SCC of 800,000 or higher. For goats, 1 million is common, and 2 million isn't unheard of. To repeat: somatic cell counts in goats apparently don't mean the same as SCCs in cows.

For commercial goat dairies, the problem becomes particularly acute in the fall and winter. Then SCCs climb, perhaps due to estrus or late lactation. Many commercial dairies strive for out-of-season breeding, specifically to have early lactation (and low SCC) milk in the fall and winter, to keep the herd SCC within allowable limits — limits based largely on what is known about cows, not goats.

In cows, high SCCs also decrease the amount of cheese a given quantity of milk yeilds. Not so with goats.

With an increase in commercial goat dairying, this problem will undoubtedly receive more attention. As of 1999, studies were being conducted at the University of Connecticut and the French National Institute of Agricultural Research. So far, the mystery has only deepened. The French researchers found that infusing udders with antibiotices when does were dried off lowered SCCs in the next lactation, but only for 75 days. After that it made no difference.

And teat dipping — a proven practice with cows and a time-honored one with goats — did not affect the SCC in goats at all. (It's still important for mastitis control, however.)

D. The Composition of Milk

"Goat milk is richer than cow milk, isn't it?" If we assume "richer" means higher in fat, the answer is "sometimes."

There is a great deal of variation in the composition of milk, not only among species but also breeds, families, and even the same individuals at different ages and stages of lactation or on different feeds.

If you look at one of those charts giving the "average" composition of milk (and seldom do two of them agree) here's what you might find:

Average Composition of Milk in Mammals

MAMMAL	FAT	PROTEIN	LACTOSE	MINERALS	TOTAL SOLIDS
Human	3.7	1.6	7.0	0.2	12.5
Cow	4.0	3.3	5.0	0.7	13.0
Goat	4.1	3.7	4.2	0.8	12.9

Similar charts for various breeds of cows show greater differences than those indicated between cows and goats. The same is true for goats. While averages show that Nubians produce milk that's richer in fat than Saanens, these, too, are only averages. Some Saanens produce milk that's richer than the milk of some Nubians.

Some of the reasons are explained in appendix B. Further, milk composition is highly heritable. Animals of the same bloodlines are likely to have the same milk characteristics (given the same health, feed, age, state of lactation, and other factors). Also, the milk of individual animals changes drastically at different periods of time: consider colostrum.

Comparative Composition of Colostrum and Normal Milk in Cows (%)

COMPONENT	COLOSTRUM	NORMAL MILK
Water	71.7	87.0
Milk fat	3.4	4.0
Casein	4.8	2.5
Globulin & albumin	15.8	0.8
Lactose	2.5	5.0
Minerals	1.8	0.7
Total solids	28.3	13.0

While colostrum presents an extreme example, changes in milk composition do occur throughout lactation with milk during the first 2 months after freshening generally being from 0.5 to 1.5 percent lower in fat than milk from the same animal during the last 2 months of lactation.

Protein, fat, and SNF (solids not fat) generally decline with age, with SNF declining even more than fat.

Butterfat tests normally rise in the fall and drop in spring.

Other factors that can cause minor changes in the composition of milk include temperature (temperatures above 70°F and below 30°F increase the fat content) and exercise (slight exercise slightly increases fat content, while more strenuous exercise decreases fat and total output).

E. Giving Injections

How do you give a goat an injection?

The best way to learn is by watching someone else. Most veterinarians will show you how to do something as basic as this; they have more important things to do with their time and education than giving shots. (Veterinarians don't "give shots," of course, or even injections; they give parenteral medication. This just means the medication isn't administered orally.)

Injections can be intramuscular (IM), subcutaneous (SC), intraperitoneal (IP), and intravenous (IV). These names refer to where the needle goes in. *Intramuscular* means within the muscle, or meat. Generally this means a rear leg. *Subcutaneous* means under the skin. *Intraperitoneal* means within the peritoneum, or the space between the intestines and other parts of the lower gut and the internal organs such as the liver. *Intravenous* means directly into a vein and is best left to a veterinarian.

Different medications require different types of injections. Read all labels carefully, and not only to learn how the medication should be administered. Note also any safety precautions, how the medicine should be stored, shelf life, withdrawal time, and any other information provided.

There are three commonly used types of syringes: all glass, glass and metal, and plastic. Most farms use disposable plastic syringes, as these don't require sterilization equipment. (*Note:* Equipment can be sterilized in a domestic pressure cooker or canner by placing it in steam under pressure for 30 minutes. Remove metal plungers so they do not expand and break the syringe. Letting syringes stand in a pan of boiling water for 30 minutes is another method, but it's not totally safe.) Disposable syringes come in sterile packages, are used once, and discarded.

There are also reusable needles, which require sterilization, and disposable needles. These come in various lengths and gauges. For goats, 1-inch needles in 16-, 18-, and 20-gauge sizes are commonly used. The larger the gauge, the smaller the needle. The choice usually depends on the medication being given, although it is also largely a matter of personal preference.

There are several ways to fill the syringe. Here is one method.

1. Wipe the rubber bottle cap with alcohol or disinfectant.
2. Place a sterile needle on a sterile syringe, pull back the plunger, and fill the syringe with air.
3. Insert the needle through the rubber cap, invert the bottle, and blow a little air into the bottle by depressing the plunger.
4. Repeat blowing in air, depressing the plunger, and withdrawing fluid until the syringe holds the desired amount.

It's not possible to sterilize the skin of a live animal, but it should be as clean as possible. A wide variety of skin disinfectants are available, including a 70 percent solution of ethyl alcohol in water, and iodine, including tincture of iodine and Lugol's iodine.

Most vaccines and antisera are given sub-cutaneously. The needle is inserted through the skin, and the substance is deposited beneath the skin.

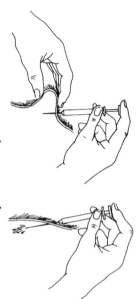

1. Grasp the skin between the thumb and forefinger and raise it into a "tent" or bubble.

2. Insert the needle at the base of this tent, ▶ parallel to the skin. Release the skin when the needle is in.

3. Push the plunger of the syringe to force ▶ the substance under the skin.

4. Withdraw the needle and syringe.

Follow the manufacturer's instructions exactly when using and storing vaccines and medications.

F. A Goat-Keeping Calendar: Plotting Tasks for a Year

January

Set up your record keeping for the year. What information and how much depends on why you keep goats and how much you enjoy keeping (and/or looking back at) records. Basic data should include production and health records to aid in culling. Include dates of worming and vaccinations and products used. Record expenses. Knowing what your milk costs is a great incentive and tool to improve your management. Record birth dates, weights, and other pertinent information, including sire and dam. (No, you will not remember these a few years from now.) If you raise registered animals or show them there will be much more.

When you order seeds for your garden, order some for a "goat garden" too! Include carrots, kale, chard, collards, and comfrey roots.

Have everything ready for kidding, including iodine, and feed pans or bottles and nipples. You might not need a heat lamp, but have one on hand, just in case. An extra bulb is good insurance.

February

If any of your does were bred in September, they'll kid in February.

Frozen water buckets or troughs can mean extra work at this time of the year, but fresh water is important.

Disbud kids before they are 2 weeks old. Castrate buck kids by 4 weeks.

February is the "make it or break it reality month," according to Sheila Nixon, who has run a commercial dairy since 1958 and grew up on a goat dairy before that. The rush season of kidding and the worst weather of the year separates the dreamers from the real goat people, she says.

Spend time with your kids. Handling and talking to them when they're young will make them easier to handle when they grow up.

March

Does bred in October will freshen this month.

How are your February kids doing? Check their progress with a monthly weigh-in. For the first 5 months they should be gaining 10 pounds a month. In other words, if the birth weight was 8 pounds, at 1 month of age the kid should weigh 18 pounds; 2 months, 28 pounds. (If you didn't weigh them at birth, figure 8 pounds as an average birth weight.) If they gain faster, good. Slower, check your management.

Check your pasture fences, and plan for any needed repairs.

If rotational grazing will work for you, lay out your paddocks now. Build the fences or order moveable fencing.

Is there an Easter or Passover market for your extra kids?

Magnesium deficiency and grass tetany can occur in early spring grazing on lush pasture, which may be high in potassium. To prevent this, feed hay first, and limit the time spent on pasture.

Thoroughly clean pens and stalls when weather permits.

April

Are there muddy spots in the goat yard? Fill, or drainage, may be called for. Are there any plants in your goat yard or pasture that might be harmful to your animals? Get a book written for your locale to identify plants you're not familiar with. Get a list of dangerous plants from your extension office.

Harvest dandelion greens as a feed supplement.

As your goats move onto pasture, watch for bloat.

Daylight savings time starts. If you're the punctual type, you can "spring ahead" on chores time about 10 minutes a day for 6 days to ease yourself and your goats into the new schedule.

May

Control flies *before* they become a problem. The number-one defense is sanitation: don't give them a place to get started. Keep bedding clean and dry, and don't let wet spots accumulate around watering devices or buckets. You might also want to use parasitic wasps, diatomaceous earth, fly paper, and traps, in any combination. Chemical sprays should be a last resort in a home dairy.

As the weather warms, scrub water buckets and troughs more frequently to eliminate scum and algae.

Clip goats to keep them cleaner and cooler, and to discourage external parasites.

June

Time to make hay, or to buy it "from the field" while neighbors are baling, so you get the best price. To assure quality and a fair price, have the hay tested. (See your county extension agent for details.)

Milk production is peaking. Use the surplus to make cheese, but also freeze some for the winter drought. If you still have too much, feed it to a calf, pig, or chickens, or cull your herd.

If you see a goat that's trembling, breathing rapidly and shallowly, and has a rising body temperature, it's probably heat exhaustion. Provide plenty of clean cool water, electrolytes, and trace minerals.

Goats can get sunburned. White-skinned animals are most suscepti-ble. Some people say that eating lush clover or buckwheat increases the chances for sunburn.

July

Hot weather means goats need more water and plenty of shade. If ani-mals are to be moved, do it at night or on cool days.

Rotate pastures as needed for optimum forage production and herd nutrition and health. Mow the weeds the goats won't eat before they go to seed, to encourage more palatable forage.

By now, February kids should weigh 50 pounds.

As the days get shorter, your buck might start to have an odor, and does might come into heat.

Be aware that rain after a long dry spell can increase the nitrate con-tent of some common pasture plants and result in poisoning.

Pinkeye becomes more common in hot, dry weather. Watch your goats to be sure their eyes don't water excessively or cloud over.

Now is the time to make goat-milk ice cream!

You might try *flushing,* or temporarily increasing energy and protein in feeds to stimulate estrus and synchronize breeding and to increase litter size.

August

This is the time to deworm and to give any needed vaccinations.

Check your production records. Decide which does will be bred early, or late, or might be milked through and which should be culled to reduce the winter feed bills.

Breed by weight, not by age. A doeling bred when she has achieved about half of her projected adult weight will be more productive, efficient, economical, and healthy than one bred later. This is usually around 80 pounds.

Watch the weather. A cold snap could start your does cycling. Record heat periods on your calendar. Prepare for a trip to the buck 16 to 18 days later for early breeding, or just keep track of the heat periods for later ref-erence. (If you don't own a buck, line one up now.)

If you plan to keep or sell doe kids, find the right buck for each doe; one that will improve your herd.

Feed the goats (buck and does) carrots for vitamin A, good greens, and a small amount of oats and bran, but cut down on legumes, which some say can affect fertility.

Be sure all fences, gates, and latches are in good condition and sturdy enough to prevent breeding accidents.

September

How about a goat barbecue for Labor Day?

As does are bred, mark the date on your calendar as well as the expected birth date (150 days later). A feed high in fiber and lower in protein than the milking ration is in order for the first 3 months of gestation.

Check your feed supply, and estimate your needs for the winter. Do you have enough hay and grain on hand? Bedding? If you lack storage space, have you locked in a regular and reliable source of quality feed?

Seed or reseed pastures.

October

Be sure your goats' housing is draft free but well ventilated. Autumn's rapid weather changes increase the potential for pneumonia. Protect your goats from drafts, and keep them dry.

Be sure your goats don't overindulge on apples or other fall produce available now. Feed these with care, and watch for bloat.

Frost can change the chemical composition of many forage plants, including Johnson grass, sorghum, sudan, and alfalfa, making them toxic.

Daylight savings time ends.

November

When you rake leaves, bag them for winter goat treats. Or if you have large quantities, use them as bedding. They will make even better compost after being in the goat barn all winter. Or use them to heavily mulch the carrots so they can be dug during the winter as a special treat, for you and the goats.

Be sure your goats get enough exercise to avoid pregnancy toxemia.

Take steps to prevent waterlines from freezing.

Check again for drafty conditions in the barn.

December

Reduce the amount of grain fed to does 4 months after breeding.

Any does still cycling? Time is running out. Pen breeding might be in order (let her run with the buck).

Inventory feed and bedding again.

Check kidding supplies.

Check and trim hooves. They grow faster when the goats are on soft bedding than when they're in rocky pastures.

Do *not* feed the discarded Christmas tree to the goats.

G. Immunizations

Every region has different health problems that can be prevented by immunizations. The person best qualified to help you set up and administer a health maintenance program is your veterinarian.

Contacting a veterinarian about vaccinations has several side benefits. It establishes contact. The doctor will know you have goats and that you care enough about them to seek professional guidance before some dire emergency arises. And a veterinarian is more likely to proffer valuable advice or assistance on other matters on a routine, nonemergency call than in a life-or-death situation.

Chances are your veterinarian will recommend the four basic vaccinations: tetanus, white-muscle disease, enterotoxemia, and pasteurellosis. (While the most prevalent goat ailment is probably caseous lymphadenitis [CL] and the AIDS-like caprine arthritis encephalitis [CAE] gets the most publicity these days, there are no vaccines for either of these. Nor are there vaccines for other viral-caused diseases, such as pneumonia and coccidiosis.)

Remember that a vaccine is not medicine, in the sense of being a "cure." It's a preventive measure. If your goat has already come down with enterotoxemia, it's too late for a vaccination.

Also keep in mind that vaccines often differ according to their manu-facturer. No book can give blanket instructions regarding these products, but even more importantly, just because you used a vaccine for a given purpose once, don't assume it will be the same next time. Always read and follow instructions on the labels carefully.

With these caveats, here are a few guidelines:

- **Enterotoxemia:** Give 5 mL SC (under the skin) to pregnant does 2 weeks before freshening. Kids get 2½ mL SC at 4 months of age.
- **White-muscle disease:** Give 2½ mL Bo-Se per 100 pounds of body weight to pregnant does SC 1 week before freshening.
- **Tetanus:** Give 1½ mL IM (intramuscular) 4 weeks before fresh-ening. This will protect young kids for disbudding and castration, but they should get a booster shot at 2 months of age.
- **Pasteurella:** Two mL IM to kids at 2 months of age. Repeat in 2 weeks.

There are other vaccinations, soremouth being one of the more prominent. But many goat owners resist vaccinating kids for soremouth because it gives kids a sometimes bad case of the disease.

Of course, there are always dangers. All injections bring the risk of anaphylactic reaction. (Anaphylaxis is sensitivity to drugs or foreign pro-teins introduced into the body resulting from sensitization following prior contact with the causative agent.) Always watch your animals for 10 minutes after any injections, and have a bottle of epinephrine on hand, just in case. And keep your veterinarian's phone number handy.

H. Metric Conversion Chart

UNIT	METRIC EQUIVALENT	ROUND EQUIVALENT
AREA		
1 square foot	0.09290304 sq m	0.09 sq m
1 acre	0.405 hectare	0.4 hectare
DISTANCE		
1 inch	2.54 cm	2.5 cm
1 foot	30.5 cm (0.305 m)	0.3 m
TEMPERATURE*		
0°F	−18°C	
32°F	0°C	
70°F	21°C	
VOLUME		
1 teaspoon	4.92892 mL	5 mL
1 tablespoon	14.7868 mL	15 mL
1 cup	236.588 mL (0.236588 L)	230 mL
1 pint	0.473176 L	0.5 L
1 quart	946.353 mL (0.946353 L)	0.95 L
1 gallon	3.78541 L	3.8 L
WEIGHT		
1 ounce	28.35 g	28 g
1 pound	453.6 g (0.45 kg)	454 g

*To convert Fahrenheit to Celsius, subtract 32 from the Fahrenheit number. Divide the answer by 9. Multiply the answer by 5.

GLOSSARY

Abomasum. The fourth or true stomach of a ruminant where enzymatic digestion occurs.

Abscess. Boil; a localized collection of pus.

ADGA. American Dairy Goat Association, the oldest and largest dairy goat registry in the United States.

Afterbirth. The placenta and associated membranes expelled from the uterus after kidding.

AGS. American Goat Society, a registry.

AI. Artificial insemination.

Alveoli. Tiny hollow spheres in the udder whose cells secrete milk. Singluar: alveolus.

American. A doe that is ⅞th purebred and recorded with ADGA; a buck that is ¹⁵⁄₁₆ths purebred and recorded with ADGA.

Anthelmintic. A drug that kills worms.

Antitrypsin factor. A substance that prevents the enzyme trypsin in pancreatic juice from helping to break down proteins. Present in soybeans.

AR (advanced registry). A designation for a goat that has produced at least 1,500 pounds of milk in a 305-day lactation.

Ash. The mineral matter of a feed; what is left after complete incineration of the organic matter.

Balling gun. Device used to administer a bolus (a large pill).

Barn records. A tally of daily milk production kept by the goat owner rather than by an official testing organization.

Blind teat. A nonfunctioning half of an udder (usually due to mastitis).

Bloat. An excessive accumulation of gas in the rumen and reticulum, resulting in distension.

Bolus. A large pill for animals; also, regurgitated food that has been chewed (cud).

Breed. Animals with similar characteristics of conformation and color, which when mated together produce offspring with the same characteristics; the mating of animals.

Breeding season. The period when goats will breed, usually from September to December.

Buck. A male goat.

Buckling. A young male.

Browse. Bushy or woody plants; to eat such plants.

Buck rag. A cloth rubbed on a buck and imbued with his odor and kept in a closed container; used by exposing to a doe and observing her reaction to help determine if she's in heat.

Burdizzo. A castrating device that crushes the spermatic cords to render a buck or buckling sterile.

Butterfat. The natural fat in milk; cream.

CAE (caprine arthritis encephalitis). A serious and widespread type of arthritis, caused by a retrovirus.

California mastitis test (CMT). A do-it-yourself kit to determine if a doe has mastitis.

Caprine. Pertaining to or derived from a goat.

Carbonaceous hay. Any hay that is not a legume (such as the clovers and alfalfa) including timothy, brome, Johnson grass, and Bermuda grass.

Chevon. Goat meat.

Cistern. Final temporary storage area of milk in the udder.

Classification. A system of scoring goats based on appearance.

Colostrum. The first thick, yellowish milk a goat produces after giving birth, rich in antibodies without which the newborn has little chance of survival.

Concentrate. The nonforage part of a goat's diet, principally grain, but including oil meal and other feed supplements, that is high in energy and low in crude fiber.

Confinement feeding. Feeding goats restricted to a barn and exercise yard, that is, nonpastured goats.

Conformation. The overall physical attributes of an animal; its shape and design.

Creep feeder. An enclosed feeder for supplementing the ration of kids, but which excludes larger animals.

Cull. To remove a substandard animal from a herd; also, such a substandard animal.

Dairy cleaning agents. Alkaline or acid detergents for washing milking equipment; iodine or chlorine compounds for sanitizing milking equipment.

Dam. Female parent.

DHIA (Dairy Herd Improvement Association). A program administered by the USDA, through Extension Services, to test and record milk production of cows and goats.

DHIR (Dairy Herd Improvement Registry). A milk production testing program administered by dairy goat registries in cooperation with DHIA.

Disbudding iron. A tool, usually electric, that is heated to burn the horn buds from young animals to prevent horn growth.

Dished face. The concave nose of the Saanen.

Doe. A female goat.

Doeling. A young female.

Drenching. Giving medication from a bottle.

Dry period. The time when a goat is not producing milk.

Drylot. An animal enclosure having no vegetation.

Elastrator rings. Castrating rings resembling rubber bands; they are applied with a special tool called an elastrator to the scrotum so it will atrophy and fall off.

Electrolyte. Mineral salts necessary for life, including sodium, potassium, calcium, and magnesium, and are lost when a body loses more fluid than it can take in.

Enterotoxemia. A bacterial infection, usually resulting in death; also called pulpy kidney disease and overeating disease.

Feed additive. Anything added to a feed, including preservatives, growth promotants and medications.

Flushing. Feeding females more generously 2 to 3 weeks before breeding in order to stimulate the onset of heat, induce the shedding of more eggs resulting in more offspring, and improving the chances of conception.

Forage. The hay and/or grassy portion of a goat's diet.

Free choice. Free to eat at will with food (especially hay) always present.

Freshen. To give birth (kid) and come into milk.

Gestation. The time between breeding and kidding (average 150 days).

Grade. A goat that is not purebred, or cannot be proven pure by registry records; any goat of mixed or unknown ancestry.

Grade A. A category of licensed dairy meeting strict regulations for equipment, milk handling, and sanitation.

Green forage. The green, growing plant component of a goat's diet.

Growthy. Description of an animal that is large and well-developed for its age.

Hand feeding. Providing a measured amount of feed at set intervals.

Hand mating. Controlled breeding, as opposed to letting a male run loose with or in a pen of unbred females.

Hay. Dried forage.

Haylage. Silage made from hay plants such as alfalfa.

Heat. Estrus; the condition of a doe being ready to breed.

Hermaphrodite. A sterile animal with reproductive organs of both sexes, generally associated with the mating of two naturally polled animals.

Homozygous. Containing either but not both members of a pair of alleles.

Hormone. A chemical secreted into the bloodstream by an endocrine gland, bringing about a physiological response in another part of the body.

Horn bud. Small bumps from which horns grow.

Intradermal. Into or between the layers of the skin.

Intraperitoneal. Within the peritoneal cavity.

Intravenous. Within a vein.

IM (intramuscular). Within the muscle.

Inbreeding. The mating of closely related individuals.

IU (international unit). A standard unit of potency of a biologic agent such as a vitamin or antibiotic.

Johne's disease. A wasting, often fatal form of enteritis.

Ketosis. Overaccumulation of ketones in the body, responsible for pregnancy disease, acetonemia, twin lambing disease, and others that occur at the end of pregnancy or within a month of kidding.

Kid. A goat under 1 year of age; to give birth.

Koumiss. A fermented goat-milk drink originally from central Asia and made of mare's milk. Also spelled kumiss.

Lactation. The period in which a goat is producing milk; the secretion or formation of milk.

Lactation curve. Daily milk production as represented on a graph, usually rapidly rising soon after freshening, then slowly falling.

Legume. A family of plants having nodules on the roots bearing nitrogen-fixing bacteria, including alfalfa and the clovers.

Linear appraisal. A system of scoring goats on individual conformation traits.

Linebreeding. A form of inbreeding that attempts to concentrate the genetic makeup of some ancestor.

Mastitis. Inflammation of the udder, usually caused by an infection.

Microorganism. Any living creature of microscopic size, especially bacteria and protozoa.

Milking bench (or stand). A raised platform, usually with a seat for the milker and a stanchion for the goat's neck, that a goat stands upon to be milked.

Milking through. Milking a goat for more than 1 year.

Milkstone. Cloudy, bacteria-inhabited film left by alkaline detergents.

New Zealand fencing. A system of electric fencing using a high-energy charger.

Off feed. Not eating as much as normal.

Out of. Mothered by.

Overconditioned. Overfed; fat.

Papers. Certificates of registration or recordation.

Pedigree. A paper showing an animal's forebears.

Pessary. A vaginal suppository, used after kidding to prevent infection if human assistance in the birth has been required.

Polled. Naturally hornless.

Precocious milker. A goat that produces milk without being bred.

Protein supplement. A feed product containing more than 20 percent protein.

Purebred. An animal whose ancestry can be traced back to the establishment of a breed through the records of a registry association.

Raw milk. Milk as it comes from the goat; unpasteurized milk.

Recorded grade. A goat, either not purebred or not verifiably purebred, that is recorded with ADGA.

Registered. A goat whose birth and ancestry is recorded by a registry association.

Rennet. An enzyme used to curdle milk and make cheese.

Retained placenta. A placenta not expelled at kidding or shortly thereafter.

Reticulum. The second compartment of the ruminant stomach.

Rotational grazing. A system for pasturing livestock by which animals are turned out on one small section of pasture at a time; prevents overgrazing and sustains and renews plant growth.

Roughage. High-fiber, low total digestible nutrient feed, consisting of coarse and bulky plants or plant parts; dry or green feed with over 18 percent crude fiber.

Rumen. The first large compartment of the stomach of a goat where cellulose is broken down.

Scours. Persistent diarrhea in young animals.

Service. Mating.

Settled. Having become pregnant.

Silage. Fodder preserved by fermentation; also called haylage.

Sire. Male parent; to father.

Soiling. Harvesting and bringing feed to goats.

Stanchion. A device for restraining a goat by the neck for feeding or milking.

Standing heat. The period during which a doe will accept a buck for mating, usually about 24 hours.

Star milker. A designation of high milk production based on a 1-day test, not the entire lactation. ★M, ★★M, and so on indicates that the dam and granddam also held ★M status. ★B or star buck indicates star milkers in a buck's family tree.

Straw. Dried plant matter (usually oat, wheat, or barley leaves and stems) used as bedding; also, the glass tube semen is stored in for AI.

Streak canal. Opening at the end of a teat, surrounded by sphincter muscles.

Strip. To remove the last milk from the udder.

Strip cup. A cup into which the first squirt of milk from each teat is directed and which will show any abnormalities that might be in the milk.

Subcutaneous. Beneath the skin.

Synthesis. The bringing together of two or more substances to form a new material.

Tattoo. Permanent identification of animals produced by placing indelible ink under the skin, generally in the ear but in the tail web of La Manchas.

Test (to be on test; official test). To have daily milk production weighed and its butterfat content determined by a person other than the goat's owner.

Therm. Unit of measurement of energy, used with animal feeds instead of calories. One therm is the amount of heat required to raise the temperature of 1,000 kilograms of water 1°C (1 therm = 1,000,000 calories).

Total digestible nutrient (TDN). The energy value of a feed.

Toxic. Of a poisonous nature.

Trace mineral. A mineral nutrient essential to animal health, but used only in very minute quantities.

Type. The combination of characteristics that makes an animal suited for a specific purpose, such as "dairy type," or "meat type."

Udder. An encased group of mammary glands provided with a teat or nipple.

Udder wash. A dilute chemical solution, usually an iodine compound, for washing udders before milking.

Unrecorded grade. A grade goat not recorded with any registry association.

Upgrade. To improve the next generation by breeding a doe to a superior buck.

Vermifuge. Any chemical substance administered to an animal to kill internal parasitic worms.

Wattle. Small, fleshy appendage. Wattles are hereditary, not all goats have them, and they serve no useful purpose.

Wether. A castrated buck.

Whey. The liquid remaining when the curd is removed from curdled milk when making cheese.

WMT (Wisconsin mastitis test). A do-it-yourself kit to determine if a doe has mastitis.

INDEX

Note: Page numbers in *italic* indicate illustrations; those in **boldface** indicate charts or tables.